THE GREAT PALEOZOIC CRISIS
LIFE AND DEATH IN THE PERMIAN

The Critical Moments in Paleobiology and Earth
History Series,
David J.Bottjer and Richard K. Bambach, Editors

Betsey Dexter Dyer and Robert Alan Obar,
Tracing the History of Eukaryotic Cells: The Enigmatic Smile

Donald R. Prothero, *The Eocene-Oligocene Transition: Paradise Lost*

Mark A.S. McMenamin and Dianna Schulte McMenamin,
The Emergence of Animals: The Cambrian Breakthrough

The Perspectives in Paleobiology and Earth History Series,
David J. Bottjer and Richard K. Bambach, Editors

Anthony Hallam,
Phanerozoic Sea-Level Changes

The Columbia University Press Advisory Committee
for Paleontology:

DAVID J. BOTTJER, CHAIR
RICHARD K. BAMBACH
DAVID L. DILCHER
NILES ELDREDGE
S. DAVID WEBB

The concept for these series was suggested by Mark and Dianna
McMenamin, whose book *The Emergence of Animals* was the
first to be published.

The Great Paleozoic Crisis:

LIFE AND DEATH IN THE PERMIAN

DOUGLAS H. ERWIN

COLUMBIA UNIVERSITY PRESS

NEW YORK

Columbia University Press
New York   Chichester, West Sussex
Copyright © 1993 Columbia University Press
All rights reserved

Library of Congress Cataloging-in-Publication Data

Erwin, Douglas H., 1958–
The great Paleozoic crisis : life and death in the Permian /
Douglas H. Erwin.
p.   cm. — (The critical moments in paleobiology and earth
history series)
Includes bibliographical references and index.
ISBN 0-231-07466-2 (acid-free).
ISBN 0-231-07467-0 (pbk. : acid-free).
1. Paleontology—Permian.   2. Extinction (Biology).   I. Title.
II. Series.
QE730.E79   1993
560'.1729–dc20   93-1915
                                                        CIP

Casebound editions of Columbia University Press books are printed on
permanent and durable acid-free paper.

Printed in the United States of America
c 10 9 8 7 6 5 4 3

CONTENTS

Preface

From my own arguably skewed perspective, there are six really important events in the history of life on earth: the origin of life 3.5 billion years ago, the origin of eukaryotes (and sex!) at least 1.5 billion years ago, the first appearance and subsequent rapid radiation of animals 550 million years ago, the invasion of land by plants and animals during the Lower Paleozoic, the mass extinction 250 million years ago at the end of the Permian, and the changes in the Cretaceous and Early Tertiary, including the appearance of flowering plants, the disappearance of the dinosaurs, and the radiation of Tertiary mammals. And I include the Cretaceous-Tertiary events only to forestall a lot of caterwauling from a few mammals. Quite a lot else happened in between, much of it quite interesting, but these are the events that provide the structure to the history of life. Not surprisingly, many of the most complex questions facing paleontologists concern these same events.

This book is a history of the fifth of these events: the mass extinction at the close of the Permian period, the most severe mass extinction in the history of life. This book is not just about what lived and what died 250 million years ago but is an attempt at geologic history on the broadest scale. Unlike the trifling (at least to my mind), if better known, mass extinction that wiped out the dinosaurs at the end of the Cretaceous, the end-Permian mass extinction was a much lengthier and more complicated affair. The extinction took place during an interval of intense physical change in the earth, and, indeed, the most exciting aspect of this extinction is attempting to sort out cause and effect and the chance correlations from the causal connections between these physical events and the extinction. This episode requires delving into subjects as disparate as the larval ecology of marine invertebrates, fluctuations in stable isotopes, the history of the earth's magnetic field during the Permian, and changes in the core-mantle boundary.

I am far from an expert in many of these subjects, but I have consulted freely with more knowledgeable colleagues. No doubt geochemists, geophysicists, and vertebrate paleontologists, among others, will quibble about the emphasis or the details I choose to include, but like all authors, I was faced with conflicts between advancing the plot of the story and delving into more specialized details.

The introduction and first chapter describe the historical context of our understanding of the extinction, and introduce the world of the Permian before the extinction, including the dominant marine taxa, the terrestrial vertebrates and plants, and establish the context of the extinction within the history of life. Chapter 2 considers in detail the stratigraphy of Permo-Triassic boundary sections and what they tell us of the extinction. Chapter 3 recounts the pattern of extinction in the marine realm, while chapter 4 considers extinction patterns on land. Chapters 5–7 concentrate on the physical events of the Permian, including the tectonic events associated with the formation of the supercontinent of Pangea and Late Permian volcanism, variations in stable isotopes, and climatic changes. Chapter 8 considers these events in light of several models describing cycles in earth history. In these scenarios the Paleozoic-Mesozoic boundary marks a shift in geophysical realms, due largely to the formation of Pangea. The many models for the extinction are evaluated in chapter 9. My interest in the extinction is not (although this doubtless seems odd) in the extinction itself, but in its impact on the history of life. Thus the book closes with a look at the effects of the extinction on the history of life, including an attempt to resolve the debate over the role of the extinction in generating the changes in the marine and terrestrial biotas between the Permian and Triassic.

One final note on the structure of the book. I have included a number of boxes to set off explanatory material that might otherwise interrupt the flow of the narrative (for example, the compleat idiot's guide to stable isotopes in chapter 7, which I wrote so *I* could understand the topic), and additional, specialized material that may be primarily of interest to specialists.

THIS RESEARCH has been supported by the National Museum of Natural History (Smithsonian Institution), the Systematic Biology Program of the National Science Foundation, and the Petroleum Research Fund administered by the American Chemical Society. This book and my earlier work on the end-Permian mass extinction would not have been possible without the assistance of many scientific colleagues, whose help and assistance I gratefully acknowledge. They include Bob Anstey, Sandy Carlson, Bill DiMichele, Mary Droser, Dave Jablonski, Dave Kidder, Bob Linsley, Jack Sepkoski, Tom Vogel, Jim Valentine, and Scott Wing. All or part of the manuscript has been read by Cheryl Adams, Dick Bambach, Dave Bottjer, Dick Grant, Conrad Labandiera, Des Maxwell, Jennifer Schubert, Pete Wagner, and Ellis Yochelson, whose comments and advice, while not always taken, were a great help. I am particularly indebted to Bill Holser for his insightful review of the manuscript. I am also grateful for the work of Ed Lugenbeel and Jonathan Director at Columbia University Press for their many improvements to my manuscript. I also benefited from the outstanding librarians at the Natural History Branch of the Smithsonian Institution libraries. Most of the illustrations were produced by my research assistant, Liz Valiulis, who endured my frequent changes with aplomb and made numerous suggestions that greatly improved the final drawings. Figures 1.1 and 1.2 were drawn by Mary Parrish. Finally, this book would not have been written without the encouragement and support of my statistically significant other, Cheryl Adams, whose love managed to keep me sane (sort of).

INTRODUCTION

THE BIRTH OF STRATIGRAPHY

EVEN IN 1796 elephants were awfully difficult to miss, a fact that Georges Cuvier used in the first convincing demonstration that species had become extinct in the geological past. The growth of geology as a scientific discipline during the late 1700s brought to light many fossil plants and animals with no known living relatives, yet most natural theologians rejected the possibility of extinction, assuming members of the fossil forms lived elsewhere, patiently awaiting discovery. For many marine animals this was a plausible (and oftentimes correct) assumption but one that was increasingly difficult to maintain for large terrestrial mammals.

Cuvier (1769–1832) was an outstanding anatomist and paleontologist at the Muséum National d'Histoire Naturelle in Paris. In the 1790s he began studying several groups of land animals that he

1

believed contained extinct representatives; chief among them were the elephants. Many fossil elephants are morphologically distinct from either the Indian or African elephant, and Cuvier, arguing that it was increasingly unlikely that other living elephant species remained to be discovered, asserted that these distinctive fossils belonged to extinct species. At a 1796 meeting in Paris Cuvier presented a paper, *On the species of living and fossil elephants*, in which he provided nearly irrefutable documentation for extinction.* This was the first convincing demonstration that *any* fossils represented extinct species. Cuvier quickly uncovered evidence of beds within the Paris basin that documented the simultaneous extinction of many species.

Cuvier maintained that these simultaneous extinctions were evidence of some localized catastrophic event that had wiped out all or most of the species within a particular region. The suddenness of the disappearances convinced Cuvier that gradual environmental change could not have been responsible; such changes would have produced a gradual pattern of migration and adjustment, and species extinctions should not have been simultaneous. Moreover, the understanding of the geological record in the early 1800s favored Cuvier's position. In his studies of the geologic history of the Alps, Elie de Beaumont, Cuvier's student, put Cuvier's theory of occasional catastrophes in the past on firmer ground. De Beaumont correlated numerous episodes of mountain building with major breaks in the fossil record, including several events we now recognize as mass extinctions, and suggested that tectonic events were behind the extinctions (Rudwick 1985b). By the middle of the nineteenth century, however, Cuvier's claims for episodic and sudden physical changes leading to mass extinctions were strikingly at odds with Charles Lyell's vision of gradual, uniformitarian change.

Lyell's *Elements of Geology* became a standard for English-speaking geologists during the mid-1800s and Cuvier's sudden cataclysms had no place in Lyell's version of earth history. Lyell argued that species extinctions resulted from changing environmental conditions and that the suddenness of the disappearance of species was more apparent than real, the product of an incomplete fossil record. Lyell argued that these discontinuities would disappear with a more complete fossil record (Hoffman 1989b), a viewpoint fully supported by

---

*Despite the recent sturm und drang over mass extinctions, there have been few scholarly studies of the historical development of the subject. This section relies heavily on the masterful work of Rudwick (1985a) and Hoffman (1989b).

Darwin and others. These contrasting views of mass extinctions—as the result either of environmental catastrophes or an artifact of an incomplete fossil record—persisted past mid-century.

The difference in approach between Cuvier and Lyell and their respective supporters developed during a period when geology was virtually synonymous with stratigraphy: working out the relationships between different rocks and ordering these units to produce a relative geologic time scale. Several decades earlier the Wernerian school, based in Bavaria, characterized rock units on the basis of their physical characteristics, such as color, minerals, texture, and the like. By the early 1800s a group of younger geologists, led by William Smith in England and Alexandre Brongniart, a colleague of Cuvier's and professor at the Muséum in Paris, recognized that specific intervals of geologic time were characterized by unique assemblages of fossils, and that these assemblages provided a better method of correlating rocks of similar ages at widely different localities.

Among the most prominent of the younger geologists was Roderick Impy Murchison, a gentleman of "independent means" and a well-married former British military officer with boundless energy and little to fill his time (Stafford 1989). In the 1820s he settled on geology as a practical way to occupy himself. Murchison rapidly made the field his own. He began collaborating with Adam Sedgwick, Woodwardian Professor of Geology at Cambridge, on the relationships among the rocks of the Welsh borderlands. Murchison, Sedgwick, and their colleagues were inventing the science of stratigraphy as they went along. At that time the basics of stratigraphic correlation that geologists today take for granted were far from universally accepted. Spurred on by his colleagues, Murchison attempted several schemes to group the rocks he was mapping into more inclusive units. Finally, in 1835 Murchison named these rocks as the Silurian System, to differentiate his approach from the rock units identified by adherents to the Wernerian school. Originally Murchison applied the term *system* to a group of similar rocks within a limited geologic region, following de Beaumont's use of the French *systèm* for a group of rocks affected by the same tectonic event. Thus, system originally applied to rocks of a similar age, but with a tectonic rather than biostratigraphic connotation. Shortly after the introduction of the Silurian Murchison was engaged in a series of disputes with Henry de la Beche, first director-general of the British Geological Survey, and out of this controversy the term *system* acquired the biostratigraphic implication it retains today (Bassett 1991). Murchison was a tireless

3

proselytizer for his Silurian System and for the concept of systems in general. Soon after proposing the Silurian Murchison prevailed upon Sedgwick to describe the Cambrian System for rocks that appeared to underlie Murchison's Silurian; later they jointly described the Devonian system (see Rudwick 1985a, Hallam 1989b).

In 1838 Sedgwick coined the term *Palaeozoic* to unite his Cambrian System and Murchison's Silurian. Sedgwick argued that the biologic changes between systems was less than Murchison claimed, and that the great biologic systems, with analogy to the *System Naturae* of Linnaeus, covered a far longer time span. He later elaborated on his views:

> Perhaps the best nomenclature of our older rocks would give the name *System* to the whole Palaeozoic series. In that case the words Cambrian, Silurian, Devonian, Carboniferous, &c. would define subordinate divisions. These collective groups appear to run together through the intervention of the groups of transition or passage, such as the "Cambro-Silurian." On this scheme the *Palaeozoic system* would include all the rocks containing the older types of organic life, such as Producta, Orthis, Trilobite, Orthoceratite, &c. And in like manner, the Secondary system would . . . contain all rocks with the secondary types of organic life, such as Ammonite, Belemnite, &c. (Sedgwick 1847:160–161)

JOHN PHILLIPS AND THE ERAS OF THE PHANEROZOIC

JOHN PHILLIPS, Keeper of the Museum at York and later professor of geology at Oxford agreed with much of Murchison's work on the Silurian and Devonian systems, but thought Murchison's reliance on a limited number of characteristic fossils in recognizing systems a bit weak. Instead, Phillips urged the use of a combination of coexisting genera.

Phillips's contributions to the development of geology have been unjustly neglected by biographers and historians of geology. He was a fascinating man who played an important role in the establishment of British geology. His parents died when he was seven and he came to be raised by his uncle, William Smith, an engineer and self-taught geologist who completed the first geologic map of England in 1815. About that time Phillips left school and joined his uncle in London to serve as his assistant. In 1824 Phillips was offered a position at the new museum at York, which he accepted, embarking on a career in geology in his own right. He participated in the formation of the British Association for the Advancement of Science in 1831 and served as an assistant secretary for many years. From his position at York Phillips went on to become professor of geology at King's College, London, at Dublin College, and finally at Oxford University. He later recollected: "Educated in no college, I have professed Geology in Three Universities, and in each have found this branch of science firmly supported by scholars, philosophers and divines" (cited in Evans 1874:xli). Phillips's success as a lecturer, in the activities of many scientific societies, and in his active research in astronomy, meteorology, and several branches of geology reflects both his industry and highly practical nature. As a paleontologist for the Geological Survey of Great Britain, Phillips gained a detailed knowledge of fossil distributions and of British geology, particularly in Yorkshire and Oxfordshire. His scientific stature was such that he and J. Kenyon Blackwell were appointed to investigate a number of coal-mine explosions and their report led to the establishment of the first system of coal-mine inspections.

In 1837 Phillips, by this time the first analytical paleontologist, criticized Murchison, arguing that his definition of systems was insufficiently comparative and comprehensive and that a more statistical perspective was required (Bassett 1991). Phillips favored reliance on the "preponderance of various zoological classes and families," and, recognizing that these fossil assemblages fell naturally into three phases, enlarged upon Sedgwick's use of system. In adopting Sedgwick's view that the organic components of a system should be as distinctive as the living biota chronicled by Linnaeus, Phillips held that distinctive sets of organisms developed during each system, and the first two systems ended with the disappearance of these characteristic faunal assemblages. Within each phase the dominant fossils were alike, but there were great differences between phases. Phillips labeled these divisions the Paleozoic ("Ancient Life"), Mesozoic ("Middle Life"), and Kainozoic ("recent life"; later Cainozoic, then Cenozoic) systems (Phillips 1840a, b, 1841) (figure 1). In discussing his new classification Phillips wrote:

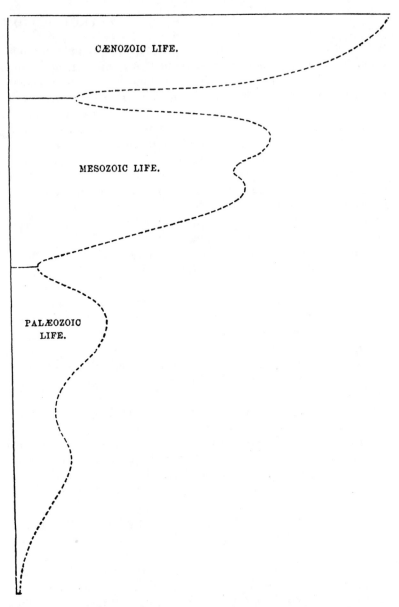

CÆNOZOIC LIFE.

MESOZOIC LIFE.

PALÆOZOIC
LIFE.

Figure 1. The original designation of the three divisions of the geologic time scale, based on fossil assemblages. The Cretaceous-Tertiary and Permo-Triassic boundaries are clearly marked, as are the Devonian mass extinction, the mid-Paleozoic diversity plateau, and the expansion in Cenozoic diversity. In attempting to circumvent the growing argument between Sedgwick and Murchison, Phillips left off both the Cambrian and Silurian. From Phillips 1860.

If . . . we resolve to employ the characters furnished by successive combinations of organic life which have appeared and vanished on the land and in the sea, we shall obtain an arrangement of remarkable simplicity, more precise in application and yet less disagreeably harsh in definition, than that which has been so long followed. We shall have three great systems of organic life, characterizable and recognizable by the prevalence of particular species, genera, families, and even orders and classes of animals and plants, but yet exhibiting, clearly and unequivocally, those *transitions* from one system of life to another, which ought to occur in every natural sequence of affinities, dependent on and coincident with a continuous succession of physical changes, which affected the atmosphere, the land, and the sea." (1841:159–160; italics in original)

Phillips's description of these three major divisions of geologic time as systems was not generally adopted, however, and Murchison's use of the term *system* prevailed. But the end of this citation reveals a view of the transitions between eras and systems shared by many geologists. These boundaries were believed to be gradational and recorded in the passage beds between systems, a view that contrasts sharply with our current emphasis on locating boundaries at a particular, although often arbitrary, horizon. There are sound reasons for the modern approach, but it does favor a more abrupt view of the change between systems.

By 1840 most British geologists recognized that the fossils of the Magnesian Limestone in England were distinct from those in the underlying Carboniferous. Not surprisingly, it was Roderick Murchison who described the Permian, eager to expand on his views of the importance of fossils to the correlation of sedimentary rocks across broad distances.

THE ORIGIN OF THE PERMIAN

IN 1840 Murchison was president of the Geological Society of London and at the height of his considerable scientific stature. That summer he visited Paris in search of more rocks to claim for his Silurian System, and learned of vast fossiliferous outcrops in Russia. He and the French paleontologist Edouard de Verneuil quickly planned a trip to Russia. Their first visit was brief but provided them with a tantalizing view of the flat-lying Paleozoic sequences deposited on the Russian platform. Their appetites whetted, in 1841 Murchison and de Verneuil returned for more intensive work, traveling first to Moscow to meet the czar and attend his son's wedding. The meeting stood them in good stead, for the czar's support allowed Murchison and de Verneuil to travel more extensively than they had the preceding summer and to consult with many Russian geologists. During their five-month journey Murchison and his companions traveled from Moscow east to the town of Perm and the Ural Mountains, south along the Urals, and then to the Sea of Azov before returning to Moscow (Murchison 1841). They found fossils characteristic of the Silurian, Devonian, and Carboniferous systems, confirming that systems recognized in England and western Europe could be correlated to deposits on the Russian platform. Murchison also investigated a series of marls, limestones, sandstones, and conglomerates containing brachiopods that were similar to those from the Carboniferous of England, and fossil fish closely related to those from the Zechstein deposits of Germany (figure 2). The Zechstein was thought to be the same age as the Magnesian Limestone in England, which in turn overlay the Carboniferous System. Furthermore, the plant fossils in these Russian rocks appeared to be intermediate between those of the Carboniferous and the Triassic.

The distinctive nature of these plant, vertebrate, and invertebrate fossils led Murchison to propose a new geologic system between the Carboniferous and Triassic. Murchison named this new system the Permian after the ancient kingdom of Permia (*not* after the small town of Perm), which had once occupied the region west of the Ural

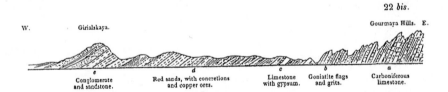

22 *bis*.

W.   Girialskaya.   Gourmaya Hills.   E.

e	*d*	*c*	*b*	*a*
Conglomerate and sandstone.	Red sands, with concretions and copper ores.	Limestone with gypsum.	Goniatite flags and grits.	Carboniferous limestone.

Figure 2. Section along the western flank of the Ural Mountains showing the conformable nature of the contact between the Carboniferous rocks (to the east) and the overlying rocks assigned by Murchison to the Permian. From Murchison, de Verneuil, and von Keyserling 1845.

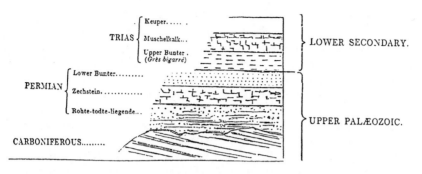

Figure 3. Diagram showing the general stratigraphy between the Paleozoic and Mesozoic (or Secondary). From Murchison, de Verneuil, and von Keyserling 1845.

Mountains (Murchison 1841; Murchison, de Verneuil, and von Keyserling 1845). In 1845 Murchison, with de Verneuil and von Keyserling, published a mammoth account of the new system entitled *The Geology of Russia in Europe and the Ural Mountains*. Rocks of equivalent age to those in Russia, Germany, and England were quickly recognized throughout the world and the Permian was soon accepted as a fundamental unit of the geologic column (figure 3). (The time units during which the systems were deposited later came to be known as periods, thus the Permian System is comprised of rocks deposited during the interval known as the Permian Period.)*

Curiously, acceptance of Murchison's Permian was more a triumph of marketing than of science, for not only had rocks of the same age been described earlier under a different name, but the earlier description was more accurate than Murchison's. In 1808 D'Omalius d'Halloy described rocks equivalent to Murchison's Permian as the *Terrain Penéen*, meaning rocks with few fossils. The sequence described by d'Halloy was well known to mining geologists in Germany, for it included the copper ore of the Zechstein deposits and a red underlayer of conglomerates, shales, and sandstones. In 1822 W. D. Conybeare and W. Phillips had correlated the Zechstein with the Magnesian Limestone and the Red Conglomerate of Devon with the Red Underlayer (now known as the Rotliegende, formerly Roth-todt-

---

*Excellent discussions of the history of the development of the geologic periods can be found in Hallam 1989a, Secord 1986, Rudwick 1985a, b, and Desmond 1984.

liegende, or red dead layers). D'Halloy's *Terrain Penéen* also included the Bunter, Muschelkalk, and Keuper groups above the Zechstein, but after von Alberti's description of the Trias in 1834 d'Halloy agreed that the later three units belonged in the Trias rather than the *Terrain Penéen*. Apparently d'Halloy lacked the authority that Murchison carried, for his work attracted little attention (Sherlock 1948). In terms of current usage, the sequence originally designated as Permian by Murchison includes rocks of Carboniferous, Upper Permian, and Triassic age. In fact, d'Halloy in 1834 was closer to our present concept of the Permian than was Murchison. In 1861 Geinitz proposed the name *Dyas* for strata included within Murchison's Permian, but Geinitz's Dyas corresponds to our modern view of the limits of the Permian system. Murchison aggressively defended the priority of his Permian, while denying that of d'Halloy. Sherlock (1948) suggests that, shorn of its Carboniferous and Triassic components, Murchison's Permian is actually equivalent to a series in modern usage. Nonetheless, Murchison carried the day with the Permian.

Murchison acknowledged that faunas within the Paleozoic, Mesozoic, or Cenozoic were more similar than faunas across the boundaries, a point also made by Phillips and Sedgwick. Murchison wrote:

> The Carboniferous and Permian fossils have that striking continuity of character . . . whilst the Permian and Triassic fossils are entirely distinct. . . . The Permian system constitutes, in fact, the remnant of the earliest creation of animals . . . and exhibits the last of the partial and successive alternations which those creatures underwent before their final disappearance. The dwindling away and extinction of many of the types produced and multiplied in such profusion during the anterior epochs, and the creation of a new class of large animals, the Saurians, clearly announce the end of the long Palaeozoic period and the beginning of a new order of zoological conditions. The two great revolutions in the extinct world are those which separated the Palaeozoic from the secondary age, and the latter from the Tertiary. Viewed as the conclusion of the first of those epochs, the Permian deposits must, therefore, excite in the minds of geologists an interest not inferior to that connected with the upper chalk, in displaying a similar apparent termination to the series of organic bodies. (Murchison, de Verneuil, and von Keyserling 1845:204–205)

Phillips elaborated his own views in *Life on the Earth* (1860), the Rede Lectures at the University of Cambridge. This is a fascinating work, an intriguing combination of modern paleobiology and ideas archaic even by the standards of 1860. *Life on the Earth* includes a prolonged defense of a divine creator and the fixity of species, as well as a surprisingly muted criticism of Darwin's *Origin of Species* (although Darwin's letters to Asa Gray and Charles Lyell indicate that Darwin took the comments very much to heart). Viewed from the present, Phillips's discussion of natural selection and descent with modification appears based more on religious conviction than science. While his religion no doubt played some role in his views, from his work as a paleontologist with the British Geological Survey Phillips was well aware that most species first appear at the base of a formation, persist through one to several formations, then disappear. Yet Phillips was no theoretician. Like Darwin he drew his conclusions from the data he knew best, in his case the fossils of Oxfordshire. As a stratigrapher Phillips could find little evidence for gradual evolutionary change during the life span of a species.

Phillips had a rather bleak view of the creator's ingenuity, however, for his creator established new species at the drop of a hat but was unable to come up with any new classes after the Cambrian, leaving the reader in a bit of a quandary. Like Murchison in 1845, Phillips claimed that the characteristic fossils of the Paleozoic, Mesozoic, and Cenozoic were produced by independent episodes of species creation, but he acknowledged that "all the important classes of Marine Invertebrates are traced into the Lower Palaeozoic Strata, beginning in each case with few species and very few genera" (Phillips 1860:79). Apparently Phillips's creator did his/her/its best work early.

The combined efforts of these early paleontologists and stratigraphers led to the relative geologic time scale by the close of the century (figure 4). But missing from the framework was any firm concept of the amount of absolute time represented. Darwin insisted upon considerable time for the evolution of life, and in *The Origin of Species* he estimated that some 300 million years had elapsed since the Cretaceous. Phillips garnered considerable enmity from Darwin for demonstrating in *Life on the Earth* that Darwin's estimates were greatly inflated (as indeed they were). When the discovery of radioactive isotopes allowed the calibration of the relative geologic time scale to absolute time in years, the 65 million years since the Cretaceous was far closer to Phillips's estimate than Darwin's.

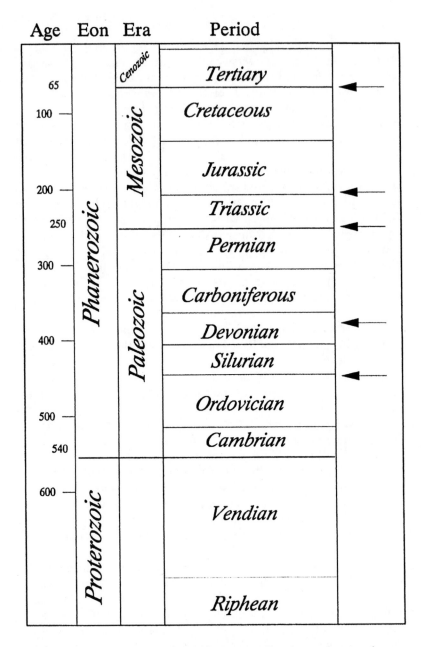

Age	Eon	Era	Period	
65		*Cenozoic*	*Tertiary*	←
100			*Cretaceous*	
200		*Mesozoic*	*Jurassic*	←
250	*Phanerozoic*		*Triassic*	←
300			*Permian*	
400		*Paleozoic*	*Carboniferous*	
			Devonian	←
			Silurian	
500			*Ordovician*	←
540			*Cambrian*	
600	*Proterozoic*		*Vendian*	
			Riphean	

Figure 4. Geologic time scale for the last 600 million years, showing the eras and periods of the Phanerozoic. The five major mass extinctions are marked by arrows in the righthand column.

MASS EXTINCTIONS

BOTH MURCHISON and Phillips exhibited a curious ambivalence about the role of mass extinctions in the history of life, an ambivalence reflected in their views on the extent of continuity and discontinuity across the Permo-Triassic boundary. This is illustrated by the two most important figures in Phillips's book of 1860. The first is his well-known graph of the diversity of life during the three eras (figure 1), which is surprisingly similar to more recent work (see chapter 1). More interesting, however, is the frontispiece to the book (figure 5) that, given the author's placement of the figure, Phillips arguably believed was the more important. Of the latter Phillips wrote:

> The system of life thus constituted in the seas of the most ancient period so far resembles the system now established in modern oceans as to contain the same classes with similar functions and dependencies. But there are great differences in the relative proportions of the classes, and of the tribes which are included in them. And these differences are to a certain extent dependent on the elapsed time; classes at first very small have grown very large, others once predominant have been greatly diminished. To make this evident it will be useful to give up the mere enumeration of species in each class, and to adopt as a basis of representation the proportions in which the classes stand to each other in each period [in other words, calculate the relative frequency]. . . . In this representation of the relative proportions of the several classes in successive geologic periods [figure 5], the arrangement is purposely made to shew by the blue tint those classes which suffer diminution with time, and by the red tint those which from small beginnings grow to great preponderance, —while the yellow tint is assigned to classes which scarcely appear in the early period, but swell out in the middle of the scale so as to equal or overmatch either of the other classes. Thus appears in a striking light the great difference between the systems of oceanic life in earlier and later periods, the nature of this difference, and something of the method of variation which binds the whole into one plan, and connects the dawn of created life with this our breathing world. (Phillips 1860:80–82)

Figure 5, the graph of relative diversity within eight classes, shows

SUCCESSIVE SYSTEMS OF MARINE INVERTEBRAL LIFE.

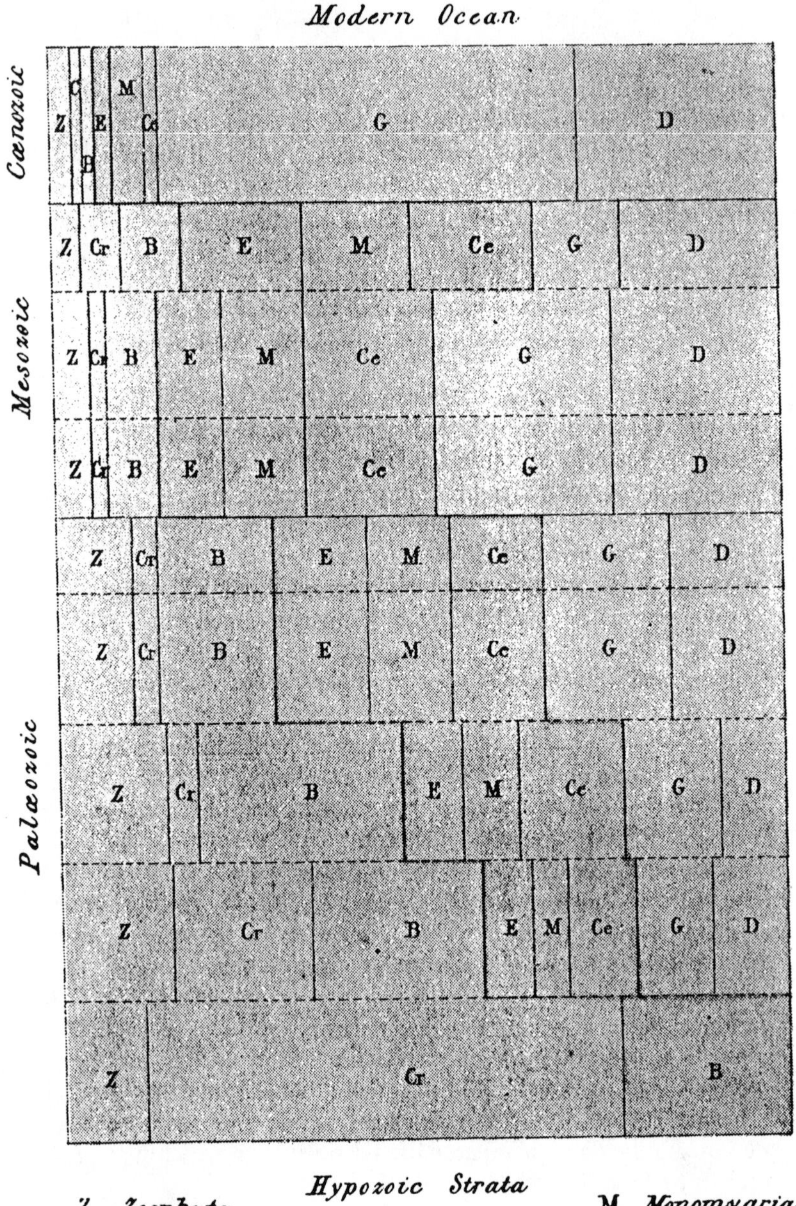

Modern Ocean

Hypozoic Strata

Z. Zoophyta
Cr. Crustacea
B. Brachiopoda
E. Echinodermata

M. Monomyaria
Cc. Cephalopoda
G. Gasteropoda
D. Dimyaria

the Mesozoic classes gradually replacing the Paleozoic classes, without any great break at the close of the Paleozoic. In a way, both figures are correct, the first illustrating changes in global species diversity and the seemingly abrupt transitions between Paleozoic, Mesozoic, and Cenozoic, the second the long-term changes in dominance between different groups. Nonetheless, the dichotomy between figures 1 and 5 neatly encapsulates the disagreement among paleontologists over the forces which control and shape the history of life.

Taken literally, figure 1 exemplifies the position of Cuvier, Sedgwick, and others, suggesting that the change from Paleozoic to Mesozoic life was largely a consequence of the mass extinction, and thus that mass extinctions are the major forces structuring biologic diversity. Charles Darwin's perspective, and to an extent Lyell's, is captured in figure 5. In this view, the structure to the history of life is provided by forces acting over long spans of time. The changes across the Permo-Triassic and Cretaceous-Tertiary boundaries may be dramatic, but at best mass extinctions merely accelerate ongoing patterns of change and do not affect the outcome.

These two different perspectives on the late Permian mass extinction persist today. Adherents to the first viewpoint argue that mass extinctions cause profound changes in the types and relative dominance of plants and animals even though the bulk of the evidence comes from the record of marine invertebrates. During the recovery following the end-Permian extinction many groups that previously had been minor components of the marine fauna diversified and became among the most important; others that suffered heavily during the extinction never reclaimed their former glory. For example, the reefs of the Permian were dominated by brachiopods, bryozoans, and stalked echinoderms, groups now regarded as important only by paleontologists. Each of these groups lived attached to the sediment, and their living descendants are relatively minor players in the modern ocean. Their place as dominant members of marine communities was taken by snails, clams, crabs, new groups of corals, and other forms. Most of these new groups burrowed into the sediment more frequently and were far more mobile; the number of predators in the fauna increased as well. Thus the composition of marine communities today in many ways reflects which groups became extinct and

Figure 5. Changes in relative diversity of eight major classes of marine invertebrates during the Phanerozoic. From Phillips 1860.

which groups survived and flourished 250 million years ago. As we will see in the final chapter of this book, some paleontologists trace this major change in the composition of marine communities directly to the end-Permian mass extinction, arguing that it removed many groups entirely, lowered the diversity of others so far that they could not recover, and allowed previously minor groups to rise to prominence. Other paleontologists do not deny the scope of the changes between the Paleozoic and Mesozoic, but have argued that most of the biological changes began long before the extinction event and, in fact, well before the Permian; thus the extinction had little impact on the scope of these changes.

Today paleontologists recognize five great mass extinctions. In order of magnitude these are the end-Permian, end-Ordovician, end-Triassic, late Devonian, and end-Cretaceous (table 1 and figure 4). More recent work suggests that there are a number of other smaller events scattered throughout the Phanerozoic, ranging from the repeated elimination of onshore trilobite assemblages in the Cambrian to a possible mass extinction in the mid-Late Miocene, but many of these extinction events are apparent only with statistical analysis, and debate continues over which are real biological events and which may be statistical artifacts. Unlike these smaller events, the five big mass extinctions all consist of a relatively rapid drop in global diversity involving several different taxonomic and ecologic groups.

The increase in the number of mass extinctions recognized by paleontologists has sharpened the continuing tension between two competing views of the evolutionary process. The first emphasizes the significance of unique events and the resulting contingent nature of history. The second perspective proposes a more law-like view of the

---

TABLE 1. Magnitude of the major mass extinctions of the Phanerozoic.

Mass Extinction	% Family Extinction	% Generic Extinction
end-Cretaceous	17	50
end-Triassic	23	48
end-Permian	57	83
late Devonian	19	50
end-Ordovician	27	57

Data from Sepkoski (1989, 1990).

history of life and suggests that we can identify overarching generalities. These are two very different ways of viewing the history of life and have profound implications for our understanding of the evolutionary process. The truth doubtless involves elements of each. If mass extinctions periodically wipe out large segments of life without respect to prior evolutionary history, adaptation and other evolutionary changes prior to a mass extinction may be irrelevant to survival during the mass extinction (Jablonski 1986a). If, however, broader evolutionary trends persist despite mass extinctions, we can have more confidence in our ability to identify the law-like generalities governing the history of life. Nowhere are these two differing interpretations more neatly juxtaposed than in the events of the end-Permian mass extinction and the subsequent Triassic recovery.

I

The World of the Permian

PHILLIPS, MURCHISON, and their colleagues realized that the plants and animals of the Permian were quite distinct from those living today. Geologists have since recognized that these organisms inhabited a world in which the distribution of the continents, the severity of storms, the oxygen content of the atmosphere, and patterns of circulation in the oceans were very different as well. This chapter introduces that world, including the animals of the Permian, the physical environment, and the extinction. The following several chapters build on this framework in discussing the Permo-Triassic boundary and the marine and terrestrial extinctions.

LIFE IN PERMIAN SEAS

EVEN THE briefest glance at an assortment of Paleozoic fossils reveals many differences from recent animals. Such familiar marine organisms as bivalves and gastropods were present in the Paleozoic but were far less important than brachiopods, bryozoans, and several stalked echinoderm groups like the crinoids. So complete was the eclipse of the latter groups during the end-Permian mass extinction that most are unfamiliar even to students of invertebrate zoology. The differences between Paleozoic and Mesozoic marine faunas are illustrated in generalized depictions of an upper Paleozoic reefal community (figure 1.1) and a Mesozoic reef to level-bottom community (figure 1.2). The Paleozoic scene is dominated by ecologically similar, but evolutionarily unrelated groups, including articulate brachiopods, stenolaemate bryozoans, and crinoids. Most were sessile, moving only during the larval stage before becoming attached to the substrate

Figure 1.1. Reconstruction of a late Paleozoic reefal community, with masses of sessile, epifaunal brachiopods and bryozoans, sponges and crinoids covering the reef mass in the right foreground. Several fish and various cephalopods are also present, but comprise a small part of the community, as do the few bivalves and gastropods.

19

Figure 1.2. Reconstruction of a late Mesozoic-level bottom community, illustrating the increase in the number and diversity of mobile taxa, including echinoids, gastropods, cephalopods, and crustaceans.

as adults. These groups were also filter-feeders, sieving food particles out of the water, or passive carnivores. Some cephalopods and fish were present but active predators comprise a fairly small part of the assemblage. Gastropods illustrate one of the most obvious differences between the Paleozoic and Mesozoic. During the Paleozoic most gastropods were herbivores, scavengers, and sessile filter-feeders but today many gastropods are sophisticated predators, often specializing on particular prey. The burrows constructed by the bivalves and other animals (in the lower left-hand corner of figure 1.1) are often shallower than their post-Paleozoic descendants. Since there was not much to eat in a brachiopod or a crinoid a Paleozoic cookbook would be thin indeed. Consequently, Paleozoic communities were capable of supporting far less diverse ecosystems than are modern marine communities.

In contrast, a late Mesozoic cookbook would more closely resemble a modern one—there was a lot more around to eat, or in scientific jargon, more available biomass. Post-Paleozoic communities could be considerably more diverse with a greater number of trophic roles—ways of making a living—and more extensive ecosystems. In figure

1.2, predators are more common and the attached, epifaunal filter-feeders of the Paleozoic have been largely replaced by mobile, infaunal and epifaunal bivalves, gastropods and crabs, along with a greater diversity of predatory cephalopods and other swimming predators.

Jack Sepkoski published a series of papers analyzing the history of marine diversity through the Phanerozoic (Sepkoski 1979, 1981, 1984), which essentially quantified the difference between the Paleozoic and post-Paleozoic faunas identified by John Phillips in 1860. Understandably, Sepkoski's work is far more rigorous and his data more extensive than that of Phillips, and, consequently, his conclusions are subtly different. Relying on the paleontological literature and on consultations with specialists in particular groups, Sepkoski arduously collected the stratigraphic ranges—the first and last appearance in the geologic column—of marine families and genera. He then applied factor analysis to the ranges of 2,800 families of fossil marine animals distributed among 91 metazoan classes. (Factor analysis is a multivariate statistical technique that reduces the variation in a large data set composed of many variables and samples to a smaller, more manageable set of composite variables or samples that explain the bulk of the variation.)

Surprisingly, Sepkoski discovered that over 90 percent of the history of marine families could be explained by what he called "three great Evolutionary Faunas" (figure 1.3). The classes within each Evolutionary Fauna share a common diversity history through the Phanerozoic, meaning that, in general, each class within a particular fauna shares a similar pattern of diversity. Classes of the Paleozoic Fauna, for example, reached their peak diversity during the Paleozoic and either have become extinct or persist with far fewer families. Similarities in familial diversity do not *necessarily* demonstrate a biological connection between the components of a single fauna. In other words, a common historical diversity pattern only suggests that independent clades respond similarly to biological and physical changes. It does not demonstrate that members of a fauna act as a cohesive biological entity. Nevertheless, the common diversity patterns within each evolutionary fauna demand further inquiry into the reasons behind their common response.

The first, and oldest, fauna includes groups that expanded rapidly during the Cambrian Metazoan radiation and dominated the Cambrian Period. This fauna is composed primarily of trilobites and inarticulate brachiopods, with lesser contributions from an odd mollusc-

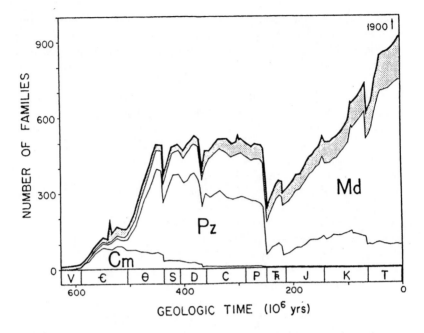

Figure 1.3. The three Evolutionary Faunas of Phanerozoic marine taxa delineated by Sepkoski's factor analysis of family diversity within classes: Cambrian fauna (Cm); Paleozoic fauna (Pz); and Modern fauna (Md) The principal groups within each fauna are shown in the following three figures. From Sepkoski (1984).

like group called hyoliths, the early molluscan monoplacophora, and some unusual echinoderms, the eocrinoids (figure 1.4). The Cambrian fauna began to dwindle in importance at the end of the Cambrian, suffered a dramatic decline during the end-Ordovician mass extinction, and quickly became an almost insignificant proportion of the marine fauna. Incidentally, Phillips recognized the distinct nature of the Cambrian Fauna and comments in *Life on the Earth* on the importance of trilobites in these faunas. By the Permian trilobites were largely restricted to reefs and were insignificant members of marine communities.

Elements of the Paleozoic Evolutionary Fauna (figure 1.5) appeared in the Cambrian but radiated primarily during the Ordovician and subsequently dominated marine ecosystems during the remainder of the Paleozoic. As mentioned above, the Paleozoic fauna is dominated

CAMBRIAN FAUNA

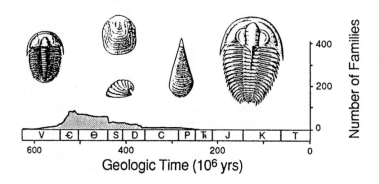

Geologic Time (10⁶ yrs)

Figure 1.4. The Cambrian Fauna is dominated by (from left to right) trilobites, inarticulate brachiopods, monoplacophorans, and hyolithids. Another trilobite is to the right. From Sepkoski (1984).

PALEOZOIC FAUNA

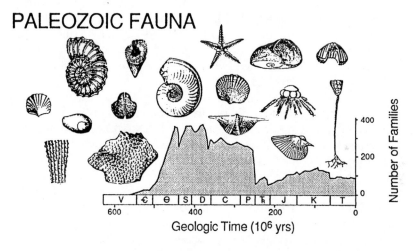

Geologic Time (10⁶ yrs)

Figure 1.5. The Paleozoic Fauna is dominated by epifaunal, sessile, filter-feeding organisms, including articulate brachiopods, stenolaemate bryozoans (lower left), rugose and tabulate corals, and crinoid and blastoid echinoderms. While the gastropods belong to the Modern Evolutionary Fauna, several groups, including the bellerophonts and sessile forms like the Macluritids (third from the left, top row) are more properly considered part of the Paleozoic Evolutionary Fauna. From Sepkoski (1984).

23

by attached, sessile, epifaunal invertebrates and suffered heavily during the end-Permian mass extinction. Prominent members of this fauna include the articulate brachiopods, the tabulate and rugose corals, stenolaemate bryozoans, and crinoid and blastoid echinoderms. Other groups include the cephalopods, starfish (the Stelleroida), and the graptolites, a group distantly related to chordates.

The bivalved shell of brachiopods lends them a superficial similarity to bivalved molluscs but they are not molluscs. Brachiopods possess a novel filter-feeding structure known as a lophophore: a fleshy, U-shaped loop bearing long tentacles that pumps water through the animal for respiration, filter-feeding, and, in some cases, direct absorption of nutrients. Most brachiopods are either free-living or attached to rocks or the bottom by a fleshy protuberance known as a pedicle. The Phylum Brachiopoda had already diversified into two classes and several orders when they first appeared at the base of the Cambrian. A tremendous radiation of articulates (Class Articulata) during the Ordovician established brachiopods as perhaps the most successful invertebrates of the Paleozoic. The group experienced an episode of almost bewildering morphologic experimentation during the Permian, particularly within the Order Strophomenida, but the class was particularly hard hit during the end-Permian extinction and has never reclaimed its former glory. Today brachiopods persist in cryptic habitats such as caves within reefs and in cooler waters such as those around New Zealand.

Bryozoans are closely related to the brachiopods. Individuals within this phylum of largely marine, colonial animals are small, but single colonies can be quite elaborate. While bryozoan colonies may resemble hydrozoans, sponges, or even calcareous algae, they are like brachiopods in that they have a lophophore. However, in bryozoans the lophophore in each individual is extended beyond the individual, much like the tentacles of a sea anemone. The delicate, frond-like colonies of the fenestrate bryozoans are common in Permian carbonate rocks, particularly in reef assemblages where they are frequently associated with articulate brachiopods, echinoderms, small corals, and some molluscs.

The third major component of Paleozoic marine communities were the stalked, or pelmatazoan, echinoderms. The phylum was of great importance during the Paleozoic; today only echinoids and starfish remain important contributors to marine communities. Chief among the pelmatazoan echinoderms were members of the Class Cri-

noidea with their flower-like cup, or calyx. A number of arms sur-
round the mouth and food is transferred down the arms to the mouth.
The calyx was generally attached to the substrate by a long, flexible
series of plates (many living crinoids are free-living).

An ecologically similar group were the tabulate and rugosan corals.
The Paleozoic coral orders were part of the Class Anthozoa, but are
only distantly related to the post-Paleozoic scleractinian corals. Tabulate corals were colonial, with long, thin corallites for the individual
polyps. Their relatives, the rugose corals, include both solitary and
colonial forms with great variation in morphology.

Brachiopods, bryozoans, pelmatazoan echinoderms, and tabulate
and rugose corals all shared similar environmental requirements,
including relatively clear water, a firm substrate and adequate nutrients. The low nutrient value of mud and the energetic expense in
removing sediment from the filter-feeding apparatus place a premium
on avoiding it. Hence filter-feeders in general avoid areas with a high
sediment flux. Firm substrate provides the attachment sites required
by many epifaunal forms. During the middle and upper Paleozoic they
were most abundant and diverse on broad, shallow shelves at lower
latitudes; in the post-Paleozoic these groups have lived in a variety
of water depths and at all latitudes.

The Modern Evolutionary Fauna originated in the Ordovician, like
the Paleozoic fauna, but expanded far more slowly in diversity (figure
1.6). The groups within this fauna were less important throughout
most of the Paleozoic than the elements of the Paleozoic Evolutionary Fauna but they diversified rapidly following the end-Permian
mass extinction. A few groups, like the ammonoids, disappeared during the Cretaceous mass extinction, but Sepkoski detected no revolution in marine faunas after that event as he did after the Permian.
The Modern Fauna dominates modern oceans and includes far more
mobile invertebrate groups than those of the Paleozoic and far more
active burrowers. The fauna also incorporates large numbers of predators and various swimming fish, reptiles, and other vertebrates.

Molluscs have been significant members of marine communities
since the Upper Cambrian, but have only become dominant since the
Permian. In Permian oceans both bivalves and gastropods were
restricted to fairly nearshore environments. Bivalves were largely
deposit feeders and filter-feeders. The class had only begun to exploit
the potential of fused siphons, which channel the inhalant and/or
exhalant water currents through the animal. Bivalves without inha-

MODERN FAUNA

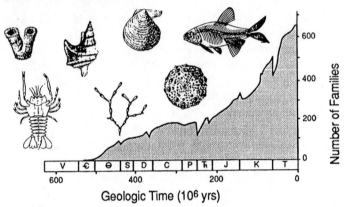

Geologic Time (10^6 yrs)

Figure 1.6. Malacostracan arthropods, scleractinian corals, gastropods, bivalves, fish echinoids, and gymnolaemate bryozoans (clockwise from left) are among the major constituents of the Modern Evolutionary Fauna. Note the fauna arose in the Ordovician and diversified steadily. From Sepkoski (1984).

lant siphons were restricted to epifaunal or very shallow infaunal depths; the development of fused siphons permitted the successful exploitation of a largely vacant niche that has since become a major bivalve habitat (Stanley 1970; Bottjer and Ausich 1986).

Traditionally, Paleozoic gastropods have been viewed as a fairly slow-moving group of grazers and filter-feeders unlike the many recent predatory gastropods. Work by a number of Paleozoic gastropod specialists reveals, however, that Paleozoic gastropods may have had an ecological range approaching that of modern gastropods, including several predatory clades (see Allmon et al. forthcoming). Since gastropods survived the mass extinction relatively unharmed (relatively is the key word here), their ecological breadth makes them an important tool in deciphering the extent of ecological selectivity in the extinction process.

Modern cephalopods largely lack a shell, but Permian oceans included two important durably skeletonized cephalopods orders: the Nautiloidea and the Ammonoidea. Both were generally coiled like the living *Nautilus* but exhibit a variety of shell forms. Cephalopods were probably one of the more important groups of predators in Permian oceans.

Arthropods, the enormous phylum that includes crabs, insects,

26

barnacles, and centipedes, also includes a number of important Paleozoic groups. Arthropods all have paired, jointed appendages and a hard exoskeleton, but the morphologic variety within the group is tremendous. Another important group of arthropods in the Permian was the ostracodes, an odd group of small (0.5–1.0 mm) crustaceans having a laterally compressed, bivalved shell, or carapace, which makes them look a bit like a seed.

The Paleozoic fauna experienced a mass extinction at the close of the Permian during which approximately 57 percent of all marine families disappeared, twice as many as disappeared during the end-Ordovician mass extinction, and perhaps as many as 95 percent of all marine species. Sessile, epifaunal filter-feeders were preferentially eliminated, while molluscs and some other groups came through relatively unscathed. The height of the extinction may have been the closest life has come to complete extermination since its origin. One might expect that a marine extinction of this magnitude would have a considerable impact on land as well, but the relationship between the marine extinction and terrestrial events is unclear. Plant floras underwent a climatically driven change as the cool, wet climates of the Permo-Carboniferous glacial regime gave way to the hotter, drier climates of the Middle and Upper Permian but this began long before the extinction. The vertebrate record is more complex. A mass extinction occurred in the Upper Permian, but this was one of a series of extinction events during the Permian and Triassic. It is always difficult to correlate between marine and terrestrial events, and thus to determine the connection between the marine and vertebrate extinctions. Yet if a connection exists, the explanation for the extinction must come from forces that affect both areas. Conversely, if there is no connection, we can eliminate from consideration mechanisms that must affect both realms.

Both Phillips and Sepkoski recognized the faunal discontinuites between the Cambrian and Ordovician and at the end of the Paleozoic, but unlike Phillips, Sepkoski did not find any major faunal discontinuity between the Mesozoic and Cenozoic, largely because he limited his study to marine families within classes, whereas Phillips, with far less data, included both terrestrial vertebrates and plants. Although it may seem obvious that the extinction was the cause of the dramatic change in the fauna across the Permo-Triassic boundary, a number of paleontologists have advanced cogent arguments to the contrary. The Modern Fauna shows a pattern of continuing, gradual expansion during the Paleozoic (figure 1.3) that suggests the classes would have come to dominate modern oceans, even without the end-Permian mass extinction. From this perspective the extinction may have speeded things up a bit but it did not substantially change the outcome. As I pointed out in the prologue, this debate goes to the heart of the significance of mass extinctions for the history of life and for the evolutionary process; we will return to the controversy in the final chapter.

Perhaps an equally interesting aspect of this debate is the possibility of a 26-million-year cycle of mass extinctions over the past 250 million years (Raup and Sepkoski 1984, 1986a). Most of these Mesozoic and Cenozoic events are fairly small and several had not previously been recognized as mass extinctions. The periodicity hypothesis implies the existence of a single, underlying causal factor tying each of these events together. Initially attention was focused on the

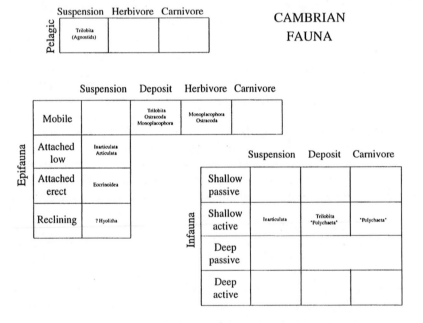

Figure 1.7.Changing patterns of resource utilization among the three Evolutionary Faunas. Both the total number of ecologic guilds occupied and the number of guilds occupied by a single class increases with successive Evolutionary Faunas. The shaded boxes are impractical ecologic strategies. From Bambach (1983).

MIDDLE AND UPPER PALEOZOIC FAUNA

Pelagic	Suspension	Herbivore	Carnivore
	Conodontophorida Graptolithina ?Cricoconarida		Cephalopoda Placodermi Merostomata Chondrichthys

Epifauna

	Suspension	Deposit	Herbivore	Carnivore
Mobile	Bivalvia	Agnatha Monoplacophora Gastropoda Ostracoda	Echinoidea Gastropoda Ostracoda Malacostraca Monoplacophora	Cephalopoda Malacostraca Stelleroidea Merostomata
Attached low	Articulata Edrioasteroida Bivalvia Inarticulata Anthozoa Stenolaemata Sclerospongia			
Attached erect	Crinoidea Anthozoa Stenoleamata Demospongia Blastoidea Cystoidea Hexactinellida			
Reclining	Articulata Hyolitha Anthozoa Stelleroidea Cricoconarida			

Infauna

	Suspension	Deposit	Carnivore
Shallow passive	Bivalvia	Rostroconchia	
Shallow active	Bivalvia Inarticulata	Trilobita Conodontophorida Bivalvia Polychaeta	Merostomata Polychaeta
Deep passive			
Deep active		Bivalvia	

MESOZOIC - CENOZOIC FAUNA

Pelagic	Suspension	Herbivore	Carnivore
	Malacostaca Gastropoda Mammalia	Osteichthyes Mammalia	Osteichthyes Chondrichthyes Mammalia Reptilia Cephalopoda

Epifauna

	Suspension	Deposit	Herbivore	Carnivore
Mobile	Bivalvia Crinoidea	Gastropoda Malacostraca	Gastropoda Polyplacophora Malacostraca Ostracoda Monoplacophora Echinoidea	Gastropoda Malacostraca Echinoidea Stelleroidea Cephalopoda
Attached low	Bivalvia Articulata Anthozoa Cirripedia Gymnolaemata Stenolaemata Polychaeta			
Attached erect	Gymnolaemata Stenoleamata Anthozoa Hexactinellida Demospongia Calcarea			
Reclining	Gastropoda Bivalvia Stelleroidea Anthozoa			

Infauna

	Suspension	Deposit	Carnivore
Shallow passive	Bivalvia Echinoidea Gastropoda	Bivalvia	Bivalvia
Shallow active	Bivalvia Polychaeta Echinoidea	Bivalvia Echinoidea Holothuroidea Polychaeta	Gastropoda Malacostraca Polychaeta
Deep passive	Bivalvia		
Deep active	Bivalvia Polychaeta Malacostraca	Bivalvia Polychaeta	Polychaeta

29

possibility of episodic impacts of comets, asteroids, or other extraterrestrial objects. The evidence for a meteorite impact at the Cretaceous-Tertiary boundary is substantial, but there is precious little evidence for such an impact associated with the end-Permian mass extinction. If Raup and Sepkoski's evidence for periodic extinctions is ever confirmed, we must then turn our search for a forcing factor inward to internal changes in the earth (a topic discussed in chapter 8). If the source of periodic extinctions *is* endogenous, however, this reopens the question of whether the Cretaceous-Tertiary extinction was in fact caused by an impact.

The ecological structure of marine communities also changed considerably during the Phanerozoic. Richard Bambach has studied changes in the occupation of various ecologic guilds through the Phanerozoic (Bambach 1983, 1985). A collection of taxa (species, genera, whatever seems appropriate) that share a common means of making a living constitute a guild. For example, all filter-feeding animals attached to the bottom constitute the epifaunal, sessile, filter-feeding guild. During the Permian this association included many different types of articulate brachiopods, bryozoans, some bivalves, several groups of gastropods, and the crinoids.

Bambach established a fairly simple guild framework based on the general adaptive strategies of marine animals (figure 1.7a–c). Two trends are immediately apparent: more guilds are occupied by each successive evolutionary fauna and, on average, more classes occupy each individual guild. Thus the complexity of ecological relationships within marine communities increases through time as does the variety of ecological roles adopted by most classes. In chapter 10 we will return to the relationship between the ecological structure of marine communities and mass extinction.

HOW GOOD IS THE RECORD OF LIFE IN ANCIENT SEAS?

JACK SEPKOSKI's diversity studies have been criticized by a number of paleontologists, most of whom object that Sepkoski and his compatriots are "casual theorists and taxon-counters," to quote the disapproving comments of one older colleague at a meeting some years ago. (Many of us took this attack as a compliment and immediately inscribed "CTTC" on our name tags.) Although Sepkoski's investigations are not without problems, I think it important to point out the enormous contribution Sepkoski's study has made to paleobiology. Before his work there was no single, comprehensive compilation of global marine diversity. Earlier paleontologists, including Otto Schindewolf, George Gaylord Simpson, Norman Newell, and James Valentine, had asked many of the same questions that Sepkoski considered but they lacked the data to attack them in the same systematic manner. More importantly, the insights gained by his results have fueled much of paleobiology for the past decade. His detractors tend to forget that the best criterion for judging a scientific contribution is the extent to which it engenders further inquiry and debate; by this criterion, Sepkoski has succeeded admirably.

Paleontologists confront several problems in studying the fossil record that are particular to studying diversity trends through time. Since this sort of data plays a major role in understanding the extinction, before relying too heavily on it we should inquire into the limitations of such studies. First, many organisms are soft-bodied and are not preserved in the fossil record; second, because each species is also a member of a genus and a family, a smaller proportion of the total number of species is likely to be preserved than representatives of genera or families; third, some apparent extinctions may be artifacts of taxonomic practice, and do not actually reflect the end of a lineage. Taking these problems in order, the number of durably skeletonized, and thus potentially fossilizable, Permian marine invertebrates was far lower than in modern oceans. We assume from this that overall marine diversity was also much lower. Today one-half to two-thirds of the invertebrate species within a marine community are unlikely to be preserved, including such ecologically important groups as annelid and polychaete worms. We know from the occasional preservation of soft-bodied faunas in the fossil record that many of these groups appeared early in the Paleozoic, but since they are preserved so infrequently we are left with little information as to their contribution to marine communities or their evolutionary history. But the assumption that the proportion of durably skeletonized to soft-bodied organisms has remained roughly constant through the Phanerozoic appears to be valid (Bambach 1977).

Ideally paleontologists would prefer a record of species diversity, but most species have restricted, local distributions and are unlikely to be both preserved in the fossil record and then recovered by paleontologists, so their record is fairly spotty. Biologists have described some 1.5 million species and estimate there may be a total of between 5 and 30 million species alive, albeit rapidly disappearing. Paleontologists have cataloged about 250,000 species of fossils, although they clearly run into the hundreds of millions, most of which were never preserved in the fossil record. But as one moves up the Linnaean taxonomic hierarchy from species to genus, family, and order, the completeness of the fossil record progressively improves. Paleontologists generally agree that the relative completeness of the fossil record for durably skeletonized marine invertebrate families yields a fair estimate of past diversity patterns, although completeness varies among groups. For example, gastropods and bivalves leave a better record than the echinoids (sea urchins), which tend to fall apart after death. Recent work suggests the fossil record for most durably skeletonized genera and families is essentially complete (Valentine 1989), although of course many of these taxa remain to be described by paleontologists. Valentine compared the fossil Pleistocene bivalves and gastropods in the Californian biotic province to the living molluscan fauna and concluded that about 85 percent of the durably skeletonized forms occur in the fossil record. The species that are absent from the fossil record are rare, small, fragile, or from deep water.

Second, while the fossil record may actually be far better for durably skeletonized species than we generally acknowledge, diversity at these levels may not directly correspond to species diversity. The total number of species appears to have increased about sevenfold during the Phanerozoic, but family diversity increased only about 2.3 times (Flessa and Jablonski 1985; see also Sepkoski et al. 1981; Bambach 1977). The progressive change in the species/family ratio is in part an unavoidable consequence of an increase in species richness and progressive branching at lower taxonomic levels (Flessa and Jablonski 1985). As the ratio of species to family has increased over time, family diversity has become a less accurate proxy for species diversity, but how important this change is remains unclear. Bambach (1989) argued that family diversity is a good proxy for species diversity at least until the Miocene. At the same time, study of the soft parts of living species has probably also had an impact on the taxonomy of younger fossil groups.

Third, there are extinctions and there are extinctions, but they are not all the same thing. Paleontologists are most interested in the disappearance of a species because all of the individuals in each population of the species has died out, disappeared, vanished. For higher taxa, all the individuals in all the populations of all the constituent species must disappear. Such terminal extinctions represent the end of a lineage. Systematists may also change the name of a lineage for a variety of reasons without any intervening extinction. These pseudoextinctions involve no "real" extinction since the lineage persists, merely under a new name (figure 1.8). If pseudoextinctions are included along with terminal extinctions, they inflate extinction records artificially. The division between the Paleozoic and Mesozoic encompasses many pseudoextinctions since paleontologists tend to specialize in the study of fossils of one or the other, but few look across the boundary.

The final set of problems that the nature of the fossil record poses for Sepkoski's analysis raises issues that must be considered in some

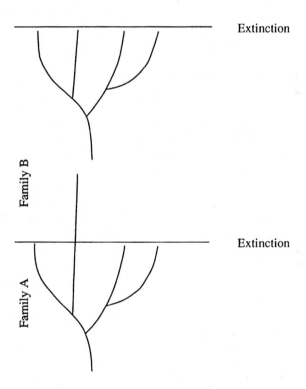

Figure 1.8. True extinctions involve termination of an entire lineage or clade, whereas a pseudoextinction reflects taxonomic practice—the lineage continues but systematists assign a new name to part of the lineage.

detail. The principal misgivings about diversity studies concern the reliability of the stratigraphic ranges and the quality and reality of the family and generic assignments. The obstacles associated with reliably gathering stratigraphic range data are numerous (e.g., Erwin 1990a). These difficulties include poor taxonomic assignments; assessing which species are correctly assigned to a particular genus, and which genera to a family; choosing between the taxonomic assignments of conflicting specialists; and accurately uncovering the stratigraphic range of the family or genus. The size of Sepkoski's database (over 4,000 families and 18,000 genera) mitigates these problems to some extent, since, if the bulk of the data is reliable, the random errors in stratigraphic ranges are unlikely to produce statistically significant patterns.

More troubling are criticisms of the taxonomic framework employed by Sepkoski. The families and genera in the database were established by systematists on a variety of grounds, and a family of ammonoid cephalopods is unlikely to be equivalent to a family of stenolaemate bryozoans. Families and genera are ranks within the nested taxonomic hierarchy established by Linnaeus, not natural groups; there are no absolute criteria to use in establishing or evaluating taxa. Consequently, views on what constitutes a family or a genus differ among specialists and often change during the career of a single specialist. Among specialists of a particular group there is commonly a shared yardstick that imparts a degree of uniformity to taxonomic assignments.

Several spirited critiques of Sepkoski's work have appeared (see particularly Smith and Patterson 1988) that suggest that quite different patterns of diversity, extinction, and radiation are revealed when a relatively new approach to systematics known as phylogenetic systematics is employed. Rather than judging taxonomic relationships by qualitative assessments of overall morphologic similarity and distinctiveness, cladistic or phylogenetic analysis emphasizes the identification of discrete morphologic characters that are shared between related taxa. The characters are employed in analyses, the objective of which is a cladogram showing the relationships among various taxa in which each taxon is descended from a single, common ancestor, or is monophyletic. For example, given three taxa A, B, and C, if A and B share more characters than B and C, A and B are more likely to be descended from a common ancestor than are A and C (figure 1.9). Characters shared by all three taxa and those that occur only in a single taxon are of no use in identifying the relationships among A,

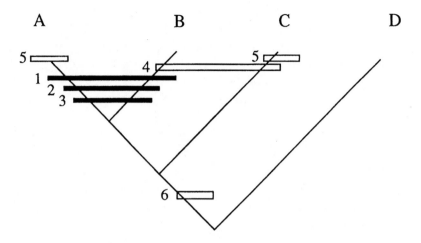

Figure 1.9. Evolutionary relationships may be reconstructed by a method of phylogenetic analysis known as parsimony. The relationships among 3 taxa (A, B, and C) are reflected in the distribution of characters among the three taxa as shown here on a cladogram. For example, if A and B share more characters (1, 2, 3) than do B and C (4) or A and C (5), A is likely to be more closely related to B than to C. Commonly the distribution of characters within such a group is compared with a close relative (D) outside the group, which may provide information on the primitive distribution of characters. In this case character 6 is shared by A, B, C, setting them off from D.

B, and C, although they may well be important in distinguishing other relationships. In practice the process is complicated because different characters may suggest different relationships. Ultimately the quality of the results of cladistic analysis depends on the quality of the characters used and which of the many available methods are used to analyze the data.

Few of the families and genera in Sepkoski's data are demonstrably monophyletic. Some are known to be polyphyletic "garbage cans," mixtures of a variety of groups descended from several different ancestors. Such groups are often established when specialists are confronted by a set of fossils they are unable to understand. A more widespread problem is paraphyletic taxa, one that includes some, but not all, of the descendants of a common ancestor. Perhaps the best example is the birds and the dinosaurs (figure 1.10). Dinosaurs have always been regarded as reptiles and placed in the Class Reptilia. Birds, given their unique feathers, flight, and other features are placed in the Class Aves. However, birds are actually descended from a group of sauris-

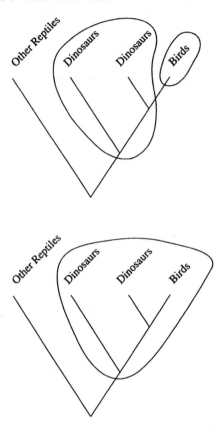

Figure 1.10. The relationships between birds and dinosaurs is a classic example of paraphyly.

chian dinosaurs known as the coelurosaurs, and the dinosaurs and the birds are more closely related than are dinosaurs and lizards. Thus, the Class Reptilia is paraphyletic because it lacks a descendant group, the birds. The solution, of course, is to place the dinosaurs and the birds in a single, monophyletic clade, known as the Class Archaeosaura.

Strictly monophyletic taxa better reflect the evolutionary history of a clade than do paraphyletic groups; there is no disagreement about the need to eliminate polyphyletic taxa. Yet every five-year-old knows that birds and dinosaurs are different, although this distinction is eliminated by a strictly monophyletic classification. Sepkoski

(1992b; also Bambach 1989, 1992) has argued that paraphyletic families or genera may better reflect changes in species diversity than strictly monophyletic taxa because of the imperfections of the fossil record. Suffice it to say that, while debate over the need for strictly monophyletic clades is far from resolved, neither this problem nor the others outlined above appears to materially affect the reality of the three evolutionary faunas. As we begin to examine the history of life on a finer scale the importance of many of these problems increases. For example, in chapter 3 we will see that the difference between monophyletic and paraphyletic clades may significantly change our view of extinction patterns during the Permian mass extinction.

TERRESTRIAL PLANTS AND VERTEBRATES

PERMIAN LAND plants and animals differed greatly from those of the Mesozoic, but many of the changes in plants occurred slowly through the Permian in response to gradual global warming. Terrestrial vertebrate patterns are more difficult to interpret, but as discussed in chapters 2 and 4 there is good evidence for a rapid turnover in composition at about the same time as the marine mass extinction. What is less clear is the connection between the marine and terrestrial turnover. Vertebrate paleontologists have identified several similar extinctions throughout the Permian and the Triassic, and there is little evidence that the extinction at the end of the Permian was substantially distinct from any of the others. Consequently, it is difficult to establish whether the terrestrial vertebrates and the marine animals were responding to related triggering mechanisms, or whether the marine and terrestrial vertebrate extinctions were unrelated. This is a critical point. If the terrestrial events were unrelated to the marine extinction the causes for the marine extinction are obviously limited to those mechanisms that affect only the marine realm.

The plant record is a bit less equivocal: as the earth's climate warmed following the end of a world-wide glaciation in the late Carboniferous and early Permian, the Paleophytic flora was replaced by a Mesophytic flora. Lower Permian plant assemblages were very similar to those in the Upper Carboniferous, but as the glaciers waned during the Lower Permian a pattern of diachronous floral change occurred (Knoll 1984). This change was marked by the decline of pteridosperms, tree ferns (pecopterids), and sphenopsids, and their replacement by conifers, callipterids, and a variety of new seed plants generally described as pteridosperms (DiMichele and Hook 1992).

Insects have a magnificent fossil record, but one that few paleontologists know much about, in part because paleoentomologists are a fairly isolated community. A growing body of work demonstrates the wealth of information available directly from the fossil record of insects, and indirectly from the damage insects left on plant material that later became fossilized. Insects continued to diversify during the Permian, particularly the herbivorous groups. Some of the modern groups were already present, including the dragonflies and mayflies of the Palaeoptera and many important groups within the Neoptera, including cockroaches, beetles, the true bugs (Hemiptera), and grasshoppers (DiMichele and Hook 1992). Conrad LaBandeira has recently demonstrated a change occurred in the dominant types of insects during the Permian, much like the change between the Paleozoic Evolutionary Fauna to the Modern Evolutionary Fauna in the marine realm, or from the Paleophytic Flora to the Mesophytic Flora. During the Permian the Palaeoptera, insects whose wings cannot be folded over the body, gave way to a modern insect fauna dominated by the Neoptera, with wings that can be folded against the body. Unfortunately the fossil record of insects in the latest Permian and earliest Triassic is spotty, so it is difficult to determine how closely these changes are associated with the marine extinction.

Permian vertebrates were a diverse group. They included cotylosaurs, pelycosaurs, and such unfamiliar groups as microsaurs, small amphibians somewhat similar to salamanders, and temnospondyl amphibians, a diverse group of terrestrial insectivores and carnivores. Cotylosaurs had an interesting mix of amphibian and amniote features. The genus *Seymouria* from the early Permian of the southwestern United States was a robust beast with generalized dentition and jaw structure. Most Permian terrestrial vertebrate assemblages were dominated by pelycosaurs, including *Dimetrodon*, with the large sail-like structure along its back. (*Dimetrodon* frequently appears in books on dinosaurs, although the pelycosaurs were only distantly related to the dinosaurs.) A variety of other reptiles also lived during the Permian.

Faunal replacement during the Permian produced a much different vertebrate fauna by the Late Permian. By this time many herbivorous tetrapods were present and assemblages were dominated by the therapsids. An extensive radiation of dicynodont therapsids occurred in the Late Permian. This herbivorous group apparently foraged close to the ground and some may have burrowed after roots and tubers.

THE FORMATION OF PANGEA

UNDOUBTEDLY THE most important geologic event in the Permian was the formation of the great supercontinent Pangea (figure 1.11). This land mass included virtually all of the large continental fragments, although some microplates never collided with Pangea. The southern continents (what are now South America, Africa, India, Antarctica, Australia, and parts of the Middle East and southeast Asia) were united into a large continent known as Gondwana during the Paleozoic. North of Gondwana lay Laurasia, which included North America, Europe, the Russian and Siberian platforms, and Kazakhstan. Reconstructing paleocontinental configurations requires careful attention to paleomagnetic and paleontologic information. Volcanic rocks record the position of the magnetic pole (declination) and the inclination of the magnetic field, which often allows reconstruction of the original paleolatitude when the rocks cooled. This information allows "paleomagicians" to reconstruct the positions of continents

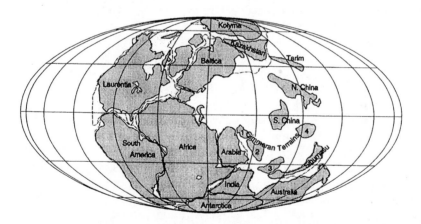

Figure 1.11. The distribution of continents at the close of the Permian about 250 million years ago. The position of the small land masses to the west is poorly constrained. Note the isolated position of the South China block. See chapter 5 for discussion.

and continental fragments through time, but the reliability of these reconstructions declines the further back in time we go. Frequently more than one reconstruction is reasonable and geologists may have insufficient data to differentiate among the alternatives.

As discussed more thoroughly in chapter 5, determining the position of the various pieces that now comprise southeast Asia is particularly demanding. Unfortunately this is also the region that contains many of the best geologic sections across the Permo-Triassic boundary. For example, most of what is now south China was a microplate floating about in the Panthalassic Ocean, a precursor to the Pacific, but the placement of this block relative to nearby landmasses is uncertain.

In the Late Carboniferous and Early Permian, Laurasia and Gondwana collided and rotated, and together the whole mass moved northward. The formation of Pangea was largely complete by the Early Permian. Indeed, microplates along the northeast margin of Gondwana rifted apart almost as soon as the supercontinent was formed. An obvious feature of the reconstructions is the large reentrant (the Tethyan seaway) along the eastern margin of Pangea.

The continental collisions involved in the formation of Pangea created a number of mountain belts. These belts, together with the supercontinent itself and changes in sea level, played a major role in Pangean climates, which will be considered further in chapter 7. The final breakup of Pangea occurred in the Triassic and Jurassic, forming the Atlantic Ocean in the process.

PERMO-CARBONIFEROUS CLIMATES

AS THE southern Gondwanan continents moved across the South Pole from the Late Devonian through the Early Permian a large continental ice sheet developed. As a result global climates were certainly cooler but, surprisingly, no major extinctions occurred, perhaps largely because the glaciation involved only a single pole. Significant global cooling seems to require formation of large continental ice sheets near both poles. The lack of any continental regions near the North Pole eliminated the possibility of substantial glaciation at high northern latitudes.

The Permo-Carboniferous glaciation ended during the Early Permian. Global climates began to warm and there is reasonably good evidence to show that the world was fairly warm by the mid-Permian. This traditional picture may be modified as a consequence of the recognition that Pangea was moving northward during the same interval. As more land neared the equator it quite naturally got warmer, a tendency that was exacerbated by the formation of Pangea. Since the interior of continents is warmer in summer (and cooler in winter) than the continental margins, because of the ameliorating effect of the oceans, the creation of a supercontinent generated extreme temperature fluctuations in the continental interior. As we will see in chapter 7, the climatic changes associated with the formation and movement of Pangea probably explains many of the changes in terrestrial plants and animals during the Permian.

The magnitude of these physical changes would in all likelihood have caused an extinction of some magnitude, although perhaps not as great as what actually occurred. The severity of the many physical events makes it difficult to identify which changes were directly responsible for the extinction, and which are just coincidence. The correlation of two events in time does not necessarily mean that there is a causal connection between them. Distinguishing between correlation and causality has always been one of the most difficult tasks in historical disciplines such as paleontology. Correspondingly, ignorance of this problem, or confusing cause and effect, frequently leads to disastrous results. The contempt of such physicists as the late Louis Alvarez toward paleontologists who were cautious in accepting the possibility that the extinction of the dinosaurs was due to an extraterrestrial impact largely reflected ignorance of the difficulties in distinguishing between correlation and causality. A review of the work on the Permian mass extinction shows that similar errors have been made in the studies of some geologists and paleontologists.

II

The Great Divide

*There are altogether 63 species, of which 17 are identical
with forms occurring already in the preceding divisions of
the productus-limestone, whilst 46 are peculiar to the sub-
division here under consideration. As far as is known to me
. . . there is not a single species which would extend into
the triassic beds of the ceratite formation . . . these beds
can barely anymore be properly called Permian; they rather
seem to form a transitional stage, a sort of passage bed
between the Palaeozoic and the mesozoic.*
— WAAGEN 1891:229–230

O NE OF THE more interesting transitions between the Permian and
Triassic occurs near Srinagar in Kashmir at a locality known as
Guryul Ravine. Here the boundary lies within a sequence of black
shale and limestone in the lowermost unit of the Khunamuh Formation. This is the sort of transitional boundary, or passage bed, that
Roderick Murchison and John Phillips expected to find between systems and that Waagen described so well in the passage cited above.
Today several geologists probably wish this locality had never been
discovered. Near the boundary the fossils include the Triassic-type
bivalve *Claraia bioni* and Permian brachiopods such as *Spinomarginifera*, combining elements of the underlying Late Permian Zewan Formation and the overlying Early Triassic of the remaining Khunamuh
Formation (figure 2.1). Murchison and Phillips would see little unusual in this bed, for systemic boundaries were a peripheral concern
to nineteenth-century geologists and they were quite comfortable

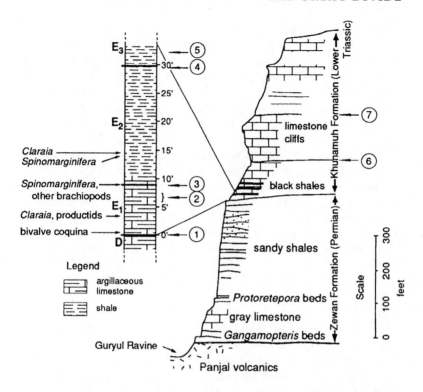

Figure 2.1. Stratigraphic section at Guryul Ravine, near Srinigar, Kashmir. The position of the Permo-Triassic boundary has been a contentious issue for most of this century. The boundary was placed at (1) by Teichert, Kummel, and Kapoor (1970), (2) by Nakazawa et al. (1975), (3) by Nakazawa (1981), (4) by Sweet (1992), (5) by Nakazawa, Bando, and Matsuda (1980), (6) by Middlemiss (1909), and (7) by Wadia (1957). Based on Kummel and Teichert (1973) with emendations from Nakazawa et al. (1975) and other sources.

with passage beds between major units. But as geology matured as a science difficulties developed in reaching international agreement on the standards and criteria to be applied in correlation. More importantly, geologists recognized that faunal or floral assemblages were historical events that often appeared at different times in different places and persisted for varying lengths of time (Newell 1978). Consequently, it was difficult to determine whether a passage bed between the Permian and Triassic in one region was precisely contemporaneous with a passage bed in another area. Since the key to stratigraphic correlation is ordering events in time, events that

43

occurred at different times in different regions, or diachronously, were of far less use than single, isochronous time horizons, like that produced by the ash from a volcanic eruption or the rapid appearance of a widespread species. Over the past century geologists have increasingly stressed boundaries between geologic units and sought isochronous layers to define them and to aid in correlation. A general stratigraphic framework for the Permian and Triassic (figure 2.2 and box 2.1) was one of the results of these endeavors.

So where should the Permo-Triassic boundary be placed in Kashmir? At the base of unit E_1 (figure 2.1) where the Triassic bivalve *Claraia* first appears (Teichert, Kummel, and Kapoor 1970)? Within the bed (Nakazawa et al. 1974)? At the top of unit E_1 after the disappearance of all the Permian-type fossils (Nakazawa and Kapoor 1981)? Or near the top of unit E_2 where a change in conodonts correlates with the top of the Permian in South China (Nakazawa, Bando, and Matsuda 1980; Sweet 1992)? Each of these positions can be defended, but in a sense they all miss the larger point, which is that the boundary between the Permian and Triassic is in some sense transitional. If either of the later two options is chosen the boundary occurs within a continuous sequence rather than at the boundary between different rock types. As we will soon see in most areas the Permo-Triassic boundary is placed at an abrupt change in rock type, or lithology (a word that should mean the study of rocks, but here refers to rock type). If a boundary occurs at a formational contact geologists rightly wonder how much time is missing. These gaps in sedimentation reduce the completeness of the record and failure to recognize them can make a gradual change appear to be an abrupt event. The apparently continuous nature of the sequence at Guryul Ravine makes it one of the most important Permo-Triassic boundary sections in the world. Studies of the Permo-Triassic boundary are replete with debates over the relative completeness of different sections. Most such debates focus on the presence or absence of gaps at contacts between different rock types, often ignoring what could be substantial intervals of nondeposition or low rates of deposition (producing a condensed section). Modern stratigraphic techniques, particularly sequence stratigraphy and studies of sedimentary completeness have not generally been applied to Permo-Triassic boundary sections, yet such studies could help resolve many important questions about the nature of the boundary.

The rate and timing of extinction is only one of many things that boundary sections reveal, albeit one of the more important. Close

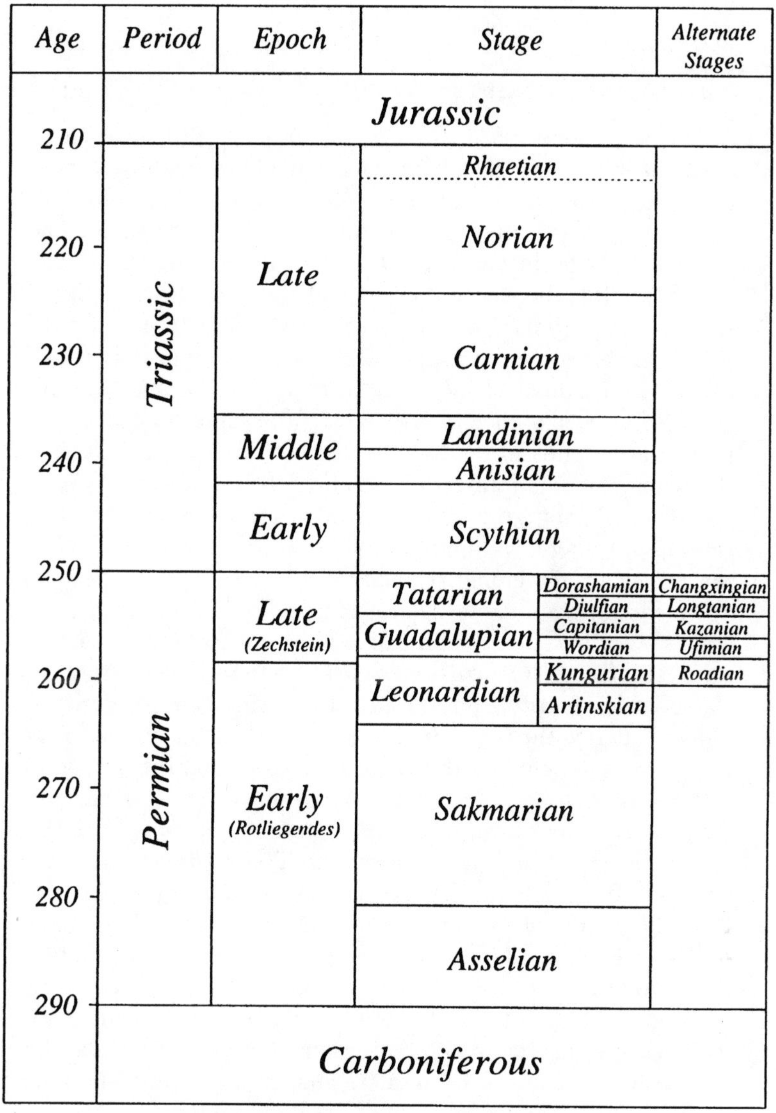

Age	Period	Epoch	Stage	Alternate Stages
210			*Jurassic*	
			Rhaetian	
220		Late	Norian	
230	*Triassic*		Carnian	
240		Middle	Landinian / Anisian	
		Early	Scythian	
250		Late (Zechstein)	Tatarian	Dorashamian / Djulfian — Changxingian / Longtanian
			Guadalupian	Capitanian / Wordian — Kazanian / Ufimian
260			Leonardian	Kungurian — Roadian / Artinskian
270	*Permian*	Early (Rotliegendes)	Sakmarian	
280				
			Assellian	
290				
			Carboniferous	

Figure 2.2. Stratigraphic chart for the Permian and Triassic. There is no general consensus on international stage-level designations, nor on how the various regional stages correlate with one another. This seems to be the most workable framework to use in this book.

45

BOX 2.1

Stratigraphy of the Permian

One requirement of history is the ability to tell time. Geologists need to order geologic events reliably to establish a relative chronology from oldest to youngest, to assign absolute dates to at least some parts of the chronology and then to be able to correlate in events that occurred in different parts of the world. Establishing these relative and absolute chronologies is the business of stratigraphy, and the trials and tribulations of this vital discipline have long been evident to geologists working in the Permian. The global stratigraphy of some geologic periods has been well established for quite some time, and current stratigraphic work in those periods involves progressive refinement of this framework. In the Permian, the problem is not regional stratigraphic frameworks, for we have plenty of those, but rather developing a global standard through correlating between the regional schemes.

Most stratigraphic work is based on groups of fossils that are widely distributed geographically, but are restricted to narrow intervals of time. Recognition of such assemblages allows a stratigrapher to identify a narrow interval of time. For example, in North America the Guadalupian Stage of the Permian Period is recognized as including the Word Limestone and the overlying Capitan Formation on the basis of distinctive brachiopods and ammonoids (Cooper and Grant 1972). Other authors note the occurrence of the distinctive ammonoid *Perrinites* in the Road Canyon Formation underneath the Word and argue that the Road Canyon is of the Guadalupian age. However, the brachiopods from the Road Canyon are typical of the older, Leonardian Stage and on this basis Cooper and Grant assigned the Road Canyon Formation to the Leonardian Stage. More recent work by Wardlaw and Grant (1991) using conodonts indicates that the boundary between the Leonardian and the Guadalupian stages may lie within the lower third of the Road Canyon.

Evidently, even in a well-studied region like West Texas—one of the major petroleum provinces in North America and the subject of continuous study since the 1930s—geologists frequently have difficulty in developing a stratigraphic framework.

Part of the problem stems from the fact that different lineages of fossils change at different times. Thus, a stratigraphic framework based on brachiopods may differ substantially from one based on ammonoids, and they both may differ from one established using conodonts.

These problems are exacerbated when correlations are attempted between different continents or between high and low latitudes. In the study of Permian rocks, the result has been the creation of a number of independent stage-level subdivisions. For the purposes of this book I will adopt the stage-level names and stratigraphic framework for the Permian and Triassic shown in figure 2.1. Not all Permian or Triassic stratigraphers may agree with these names but it is the most generally agreed upon framework.

study of boundary sections provides answers to such questions as: Was the extinction abrupt and catastrophic or were faunal disappearances spread over several million years? Was the extinction diachronous? In other words, did it begin earlier in some regions than others, or did it progress at the same rate throughout the world? This question is related in turn to issues of biogeographic pattern. Were some regions, for example the tropics, more heavily affected than others? What other patterns of selective extinction are apparent in mode of life, environment or geographic distribution? The boundary rocks may reveal evidence of marine anoxia or increased salinity, or may provide clear evidence that neither occurred, eliminating mechanisms involving such changes. In addition, comparative studies of boundary sections should reveal the extent of the marine regression. Is there any evidence for erosion at the boundary, or is the contact conformable? Boundary sections are also vital for the information they contain about stable isotopes and trace elements. Answering these questions requires high-resolution stratigraphic studies, including detailed biostratigraphic sampling through the boundary interval and analysis of stable isotopes, trace elements, magnetic polarity reversals, and sedimentology. Of course, thorough study will demonstrate that many apparently continuous sections are missing substantial pieces of the record.

The final reason for comparative analysis of boundary sections is arguably the least significant and should be of only minimal interest, although it often attracts inordinate attention from stratigraphers. Current stratigraphic practice requires selecting a particular level within a single section to serve as the boundary between the systems; the section then becomes the global reference stratotype (box 2.2) for correlation to other boundary sections. Despite the seeming parochial nature of these two issues, the position of the boundary may subtly influence our view of the extinction. Although the goal in selecting a boundary horizon is to enhance regional and global correlation, in

---

BOX 2.2

Boundary Stratotypes

The basis of modern stratigraphy, insofar as it is concerned with the boundaries between systems, is based upon the global reference stratotype. In principle, an international group of eminent specialists on the interval in question meets and agrees upon a single section of rock, and a particular level, or horizon, within that section, to serve as the global definition for, say, the base of the Triassic. By definition stratigraphers are concerned with the base of a chronostratigraphic unit rather than tops, which eliminates the potentially embarrassing situation of defining the top of unit A, the bottom of the overlying unit B, then discovering there is a pile of rock somewhere that was deposited between the top of A and the beginning of B. The global stratotype selected by our eminent committee then serves as a reference point for correlation around the world.

By convention the base of major time-rock units is fixed at a particular point in a particular section. The section where the golden spike is driven then becomes the global reference stratotype for the boundary and other sections are correlated to the stratotype.

Proposed stratotypes are judged on a number of criteria, including continuity of section, faunal content, ease of access, preservation potential, amenability to isotopic and magnetostratigraphic studies, and other characteristics.

---

fact one horizon may be promoted because it seems to emphasize a gradual extinction while another makes the extinction appear more catastrophic. Some issues that appear pivotal, such as the nature of the Permo-Triassic mixed fauna, disappear if the boundary is moved to a different horizon. In light of these difficulties I will reserve discussion of the definition of the boundary until the end of this chapter. In this chapter, as in most discussions of the boundary, discussion will focus on marine rather than terrestrial sections since the former tend to be more complete and more easily correlated than the latter.

EARLY DEFINITIONS OF THE BOUNDARY

MURCHISON'S PERMIAN section in the Urals is capped by nonmarine rocks, which, while fossiliferous, are difficult to correlate from this region (known as the type area) to anywhere else in Europe. The basal Triassic unit in the type area of Germany, the Buntsandstein, is also nonmarine, lacks useful fossils, and is likewise difficult to correlate beyond the restricted European basin in which it was deposited. Consequently the top of the marine Zechstein underlying the Buntsandstein soon came to be accepted as the Permo-Triassic boundary. During the 1860s attention shifted south to the Dolomite Alps where the marine Werfen Formation overlies the marine Bellerophon Formation. Since this boundary was thought to be at least roughly correlative with the Buntsandstein-Zechstein contact geologists accepted it as the Permo-Triassic boundary (Assereto et al. 1973; Tozer 1988a, b). The Bellerophon Formation contains a diverse Permian fauna while the overlying Werfen contains a distinctive Triassic bivalve known as *Claraia* and other Triassic fossils.

G. L. Griesbach's work in the Himalayas during the 1870s led to the discovery that the ammonoid *Otoceras* occurred in association with *Claraia* in rocks above beds of demonstrably Permian age and below Triassic fossils. Griesbach argued that the presence of *Claraia* supported correlation with the Werfen in the Alps, and that this assemblage marked the earliest Triassic through the Tethyan region. Other ammonoids also placed in the genus *Otoceras* but morphologically quite distinct from Griesbach's were known from Permian rocks in Armenia. Considerable discussion ensued over the age of *Otoceras*, with some geologists arguing for a Permian age, others for a Triassic date. Gradually it became clear that while *Otoceras*

49

occurred in both the Permian and Triassic, morphologic differences between the Armenian and Himalayan specimens made it easy to distinguish between them. The Armenian *Otoceras* were arguably older than the Himalayan species. This eliminated the correlation problem (Mojsisovics, von Waagen, and Diener 1895) and Karl Diener's (1912) demonstration of a basal Triassic age for Himalayan *Otoceras* finally set the matter to rest with the first appearance of the ammonoid genus taken as the base of the Triassic system and thus the Paleozoic-Mesozoic boundary (Tozer 1988b; the problems associated with the genus *Otoceras* are considered at length by Kummel 1972 and Nakazawa, Bando, and Matsuda 1980). Subsequently, interest in the boundary languished until new investigations began in the 1960s.

In 1961 Curt Teichert and Bernhard Kummel undertook a comprehensive reexamination of all available Permo-Triassic boundary sections. Besides a little "geotourism," they recognized that the boundary sections were critical to a better understanding of the mass extinction (Kummel and Teichert 1966). After reviewing the literature, they identified outcrops in southern China, Kashmir, northern Pakistan (the Salt and Trans-Indus ranges), the northern Iran-Azerbaijan-Armenia border region, and northeast Greenland (Kummel and Teichert 1966, 1970b) as being those most likely to have complete sequences across the boundary. They subsequently visited all of these areas except China and their contributions to our understanding of the boundary cleared up numerous problems and stimulated additional work by geologists in many countries.

MARINE BOUNDARY SECTIONS

DETAILED REVIEWS and comparisons of Permo-Triassic boundary sections have appeared (Logan and Hills 1973; Cassinis 1988; Kozur 1989; Nakazawa, Bando, and Matsuda 1980; Sweet et al., 1992; Teichert 1990; Tozer 1988b) in addition to the papers on specific sections noted below. I will highlight the nature of each boundary section, emphasizing extinction patterns, the nature of the contact, the presence or absence of a mixed-fauna, and other details that bear on extinction mechanisms (see figure 2.3 for the location of these sections during the Permian).

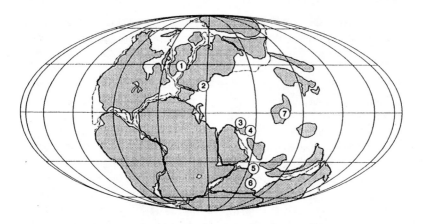

Figure 2.3. Locality map of important Permo-Triassic boundary sections discussed in the text. (1) East Greenland, (2) Southern Alps, (3) Armenia and Iran, (4) Central Iran, (5) Salt Range, Pakistan, (6) Kashmir, India, (7) South China (numerous sections).

Greenland

Rocks of Permo-Triassic age in the Kap Stosch area of central East Greenland were first reported in the late 1920s and received considerable attention in the ensuing decades, culminating in the monograph by Teichert and Kummel (1976; see also Teichert 1990). The youngest Permian unit in eastern Greenland is the Schuchert Dal Formation, a complex, intergrading sequence of sandstone, shale, and limestone representing the final filling of the East Greenland Basin. Unfortunately the age of this unit remains uncertain. Brachiopods are the most abundant Permian fossils but are no younger than Guadalupian in age (Dunbar 1955), while the presence of the ammonoid *Cyclolobus* is of little help since no consensus has been reached on the age of this genus. Teichert and Kummel (1976) judged that no latest Permian was preserved in East Greenland. More recent work in the area by teams from the Greenland Geological Survey is reviewed by Stemmerik and Piasecki (1991), who conclude that the Schuchert Dal Formation extends well into the latest Permian (Tatarian). However, they provide little biostratigraphic age control beyond that considered by Teichert and Kummel (1976). An apparently continuous Permo-Triassic section was reported from the southwestern part of the basin by Piasecki and Marcussen (1986).

East Greenland was inundated during the earliest Triassic. The shales, sandstones, and conglomerates deposited during this transgression contain *Otoceras woodwardi* and an interesting assemblage of Permian fossils, including bryozoans, brachiopods, and lesser numbers of rugose corals, ostracodes, echinoids, and crinoid stems. This mixed fauna could represent either erosion and redeposition of the Permian fossils, persistence of some Permian species beyond the boundary, or the appearance of some Triassic forms during the very latest Permian, in which case these beds should be considered Permian rather than Triassic. We will return to this subject later, but in East Greenland the Permian fossils are all representatives of species found in the underlying formations. The larger brachiopods and corals have been weathered while smaller fossils appear to have been transported within mud balls, suggesting erosion and redeposition in Triassic sediments. Mixed faunal assemblages elsewhere in the world are limited to 1–3 m above the boundary while in East Greenland Permian fossils are common through the lower 15–20 m of the Triassic, and may appear up to 100 m into the Triassic (Teichert and Kummel 1976). This suggests that at least at this locality the Permian fossils are a taphonomic artifact, not a true reflection of prolonged persistence of Permian faunas into the Triassic.

Southern Alps

Although the Alpine sections were the first region where a wholly marine Permo-Triassic contact was identified, geologists have continued to debate the continuity of the sections. In a definitive review Assereto and colleagues (1973) concluded that there is a geologic hiatus of uncertain duration throughout the region. Subsequently, the search for complete boundary sections turned eastward into the Tethyan regions of Iran, Pakistan, India, and China. But by 1986 the boundary between the Permian Bellerophon Formation and the overlying Werfen was once more considered fairly continuous (figure 2.4; Broglio Loriga et al. 1988), a view seemingly confirmed by the gradual shift in the carbon isotope record across the boundary (Holser et al. 1989, 1991).

During the late Permian the region from the Swiss-Italian border through Lombardy and the Dolomites in western Italy to southern Austria was a small, marginal Tethyan basin. This shallow inland sea deepened to the east. The Val Gardena Sandstone, a continental clas-

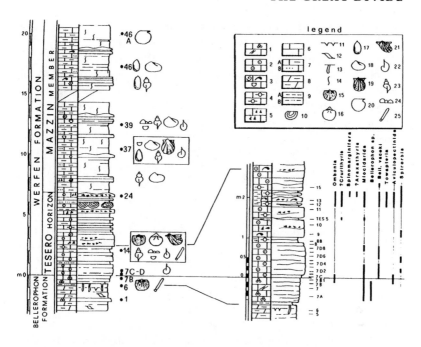

Figure 2.4. Stratigraphic section through the upper Bellerophon Formation and the Tesero Horizon and the Mazzin Member of the Werfen Formation in the Southern Alps. The Bellerophon-Werfen contact is expanded to the right, showing the stratigraphic ranges of important Permian fossils in the lower Tesero. Legend: 1) bioclastic limestone with foraminifera and algae; 2) oolitic limestone; 3) bioclastic limestone with microgastropods, bivalves, and ostracodes; 4) peloid limestone; 5) intraclastic limestone; 6) micritic limestone; 7) marly (a) and silty (b) limestone; 8) marly dolomite; 9) marl (a) and silty marl (b); 10) load structures; 11) mud cracks; 12) fenestrae; 13) vertical burrows; 14) bioturbation; 15) *Ombonia*; 16) ?*Crurithyris*; 17) *Lingula* sp.; 18) *Unionites*; 19) aviculopectinids; 20) *Claraia* gr. *wangi-griesbachi*; 21) *Towapteria*; 22) *Bellerophon vaceki*; 23) microgastropods; 24) *Spirorbis*; 25) echinoid remains. From Broglio Loriga et al. (1988).

tic deposit of red beds, sandy conglomerates, and sandstone interfingers to the east with the marine Bellerophon Formation. In the past the Bellerophon was thought to contain a transitional carbonate-evaporite unit known as the Fiammazza Facies that was replaced to the east by a shallow-water carbonate facies (the Badiota Facies)

(Assereto et al. 1973). However, detailed field studies have revealed a complex pattern of facies relationships (Cassinis 1988; Noè 1987; Broglio Loriga and Cassinis 1992). Together the Val Gardena and Bellerophon formations represent a transgressive sequence, beginning with basal continental conglomerates topped by a restricted marine lagoonal facies containing evaporites and ending in normal marine carbonates. Superimposed on the larger cycle are small-scale transgressive-regressive pulses (Broglio Loriga et al. 1988). Faunas within the Bellerophon Formation also display the transition from restricted to normal marine environments. The restricted marine fauna is limited to a few euryhaline forms, particularly algae, some foraminifera and ostracodes. A diverse marine fauna with bivalves, gastropods, crinoids, and brachiopods occurs in the normal marine facies.

The Bellerophon Formation is conformably overlain by the Werfen Formation, although whether any temporal gap exists between the formations is a subject of dispute. The basal unit of the Werfen is the Tesero Horizon, a thin, widespread oolitic unit deposited in a near-shore setting. The contact between the Tesero and the Bellerophon is gradational at most localities, although some sections do show considerable erosion (Noè 1987; Kozur 1989). The overlying Mazzin Member of the Werfen Formation is a marly mudstone with thin storm layers.

The nature of the Bellerophon-Werfen contact and the position of the Permian-Triassic boundary remains contentious. Assereto et al (1973:192) concurred with earlier studies that the contact "very closely approximates the Permian-Triassic boundary." Their correlation to sections in Armenia and Iran includes a gap of unknown duration at the top of the Permian. Broglio Loriga et al. (1988; see also Broglio Loriga and Cassinis 1992) rejected this view, based on detailed field investigations, arguing that the mixed faunal assemblage within the lower 0.5–2 m of the Tesero demonstrated a fairly continuous sequence with the boundary somewhere in the lower Tesero Horizon. Earlier geologists viewed the Permian fossils within the Tesero as reworked (Assereto et al. 1973), but unlike the situation in East Greenland, this no longer seems tenable. Broglio Loriga et al. (1988) did accept the Bellerophon-Werfen contact as the systemic boundary. While many Permian taxa do disappear with the facies change into the Tesero, others persist into the lower third of the unit (figure 2.5). This lower mixed faunal assemblage is similar to the Chinese mixed fauna bed no. 1, which also contains *Otoceras woodwardi*, leading

Buggisch and Noè (1988; see also Noè 1987) to place the Permo-Triassic boundary within the Tesero Horizon. A carbon isotope minima occurs at the same level (see chapter 7). The basal Tesero displays similar faunal and isotopic abundance patterns as the Changxing Formation. If the nearshore Tesero Horizon marks the low point of the latest Permian regression, placing the boundary within the Tesero makes intuitive sense.

Another alternative, that the boundary between the Bellerophon Formation and the Tesero is diachronous, was advanced by Kozur (1989), who claims the Bellerophon-Tesero contact in the center of the basin is equivalent to the lower portion of the Tesero on the margins of the basin. Moreover, Kozur argues that the fauna of the Tesero Horizon is equivalent to the Lower Changxingian fauna and places the entire Tesero Horizon in the Permian, with the boundary between the Tesero and the Mazzin Member of the Werfen. Kozur's argument for a diachronous boundary may bear further investigation, but on balance the detailed distribution of faunal elements within the Tesero described by Buggisch and Noè (1988) supports placement of the boundary within the Tesero Horizon.

One of the more curious features of this region is the striking increase in the abundance of fungal spores in the uppermost Bellerophon and in the Tesero Horizon. Fungal spores occur throughout the Phanerozoic but a spike this prominent is unusual. The spores and other fungal debris fragments belong to the genus *Tympanicysta* (Visscher and Brugman 1988) and have also has been recorded in Greenland, the Zechstein Basin, and throughout Tethys (Visscher and Brugman 1988; Eshet 1992). Clearly, terrestrial input into the marine realm increased but the broad geographic distribution of the spike suggests the collapse and decay of terrestrial ecosystems in the latest Permian. (The more significant recent references on these sections are Assereto et al. 1973; Broglio Loriga and Cassinis 1992; Holser and Schonlaub 1991; Kozur 1989; Noè 1987; and papers in Cassinis 1988, particularly Broglio Loriga et al. 1988. Citations to the extensive older literature may be found in these papers.) Boundary sections have also been reported from other parts of Europe and the eastern Mediterranean, including Greece (Nakazawa et al. 1975; Grant et al. 1991), Yugoslavia (Pantic-Prodanovic 1989/1990; Buser et al. 1988; Pesic et al. 1988; Broglio Loriga and Cassinis 1992), Hungary (Haas et al. 1988), Turkey (Marcoux and Baud 1988), and Israel (Hirsch and Weissbrod 1988; Eshet 1992).

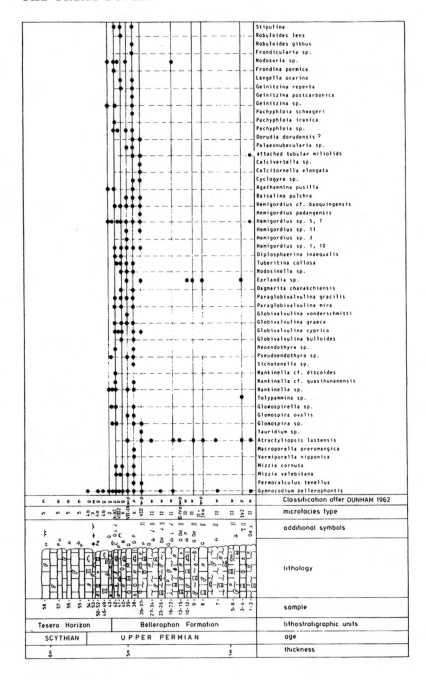

microfossils	diagenetic features	Classification after DUNHAM (1962)
ℋ algae	Q authigenic quartz	m mudstone
⌒ algal laminites	P authigenic pyrite	w wackestone
φ foraminifera	Fe limonite	p packstone
⬭ crinoids	Dd dedolomite	g grainstone
★ echinoid spines		
⊤ brachiopods	**epigenetic structures**	**lithology**
◊ gastropods	⇊ neomorphism	▭ limestone
⊟ pelecypods	↳ vuggy porosity	▭ marly limestone
～ tempestites	↟ shelter porosity	▭ marlstone
∅ ostracods	⟋ channel porosity	▭ dolomitic limestone
◇ radiolarians	↥ syntaxial overgrowth	▭ calcareous dolomite
	⌢ dog-tooth cement	▭ dolomite
macrofossils	⌣ crystal silt	▭ siltstone
C Comelicania	⊟ dissolution - compaction	▭ rauhwacke
B Bellerophon	∫⌢ stylolites	▭ stylobedding
		▭ subaquatic slides
non-skeletal particles	**penecontemporaneous deformation structures**	
⬥ peloids	⊸ birdseyes, fenestral fabric (gen.)	
⬦ pseudopeloids	⊽ load casts	
8 grapestones, lumps	⋎ mud cracks	
Ⴖ oncoids		
◯ ooids (gen.)	**organism - produced markings**	
⊗ multiple ooids	∫ slightly burrowed	
ⵣ intraclasts	∬ moderately burrowed	
⋮ detrital quartz	⊹ vertical burrows	

Figure 2.5.   Fossil distribution within the upper Bellerophon Formation and the Tesero Horizon of the Werfen Formation. From Buggisch and Noè (1988).

Transcaucasia and Iran

Studies of Permo-Triassic rocks of the Transcaucus region began with Abich's (1878) discovery of rocks he believed were of Carboniferous age. Mojsisovics (1879) restudied Abich's collections, particularly the ammonoids, and assigned them to the Permian. These ammonoids became the focus of later discussions on the Permian and Triassic boundary by Mojsisovics, Greisbach, and Diener (a detailed history of work in this area is presented by Shevyrev in Ruzhentsev and Sarycheva 1965; Teichert, Kummel, and Sweet 1973). Considerable uncertainty over the stratigraphy of the Permian and Triassic deposits and the position of the boundary persisted until Ruzhentsev and Sarycheva (1965). Following Schenck et al. (1941), Ruzhentsev and Sar-

57

ycheva divided the highly fossiliferous Upper Permian rocks into a lower Guadalupian Stage and an upper Djulfian Stage; overlying rocks were assigned to the Induan Stage of the lowermost Triassic.

The Permian and Triassic rocks form a broad syncline on either side of the Aras (or Araxes) River near the Armenian town of Dzhulfa and the adjacent town of Julfa in far northwestern Iran. The river forms the border between the two countries but the sections on either side are virtually identical, so they will be considered together, with the section at Kuh-eh-Ali-Bashi on the Iranian side serving as our example.

Teichert and Kummel established the Ali Bashi Formation for a sequence of shales and limestones that cap the Permian sequence (figure 2.6). This formation contains over 40 genera. A 4-m-thick limestone member in the upper portion of the formation contains abundant *Paratirolites*, a distinctive ammonoid. This limestone has become known as the *Paratirolites* limestone. Stepanov, Golshani, and Stocklin (1969) describe the same sequence of rocks as "Permian-Eotriassic transition beds" and, on the basis of the ammonoids, assign this approximately 18-m unit to the basal Triassic. Teichert, Kummel, and Sweet (1973) reviewed evidence from ammonoids and conodonts and concluded the unit was clearly Late Permian, as did Chao (1965) and Rostovtsev and Azaryan (1973). Moreover, Rostovtsev and Azaryan argued that the ammonoids of this unit were quite distinct from those below and erected the Dorashamian Stage above the Djulfian Stage. Sokratov (1983) reasserted (albeit unconvincingly) an earliest Triassic age for these beds. Sweet's recent conodont work (1992) confirms assignment of the Ali Bashi Formation to the Permian. Overlying the Ali-Bashi Formation are the thin-bedded gray limestones of the Elikah Formation that contain *Claraia* and other distinctive earliest Triassic fossils. This stratigraphic dispute illustrates the transitional nature of Permo-Triassic faunas and the difficulties in drawing a line with Permian rocks below and Triassic above. A detailed survey of the fauna and the stratigraphy from sections in the former Armenian SSR and the Nakhichevan ASSR was published by Ruzhentsev and Sarycheva (1965), and reviewed by Rostovtsev and Azaryan (1973). Members of the Geological Survey of Iran published the results of studies on their side of the border in Stepanov, Golshani, and Stocklin (1969); Teichert, Kummel, and Sweet (1973) examined the same sections, while Zakharov (1992) discusses a number of other sections in the Transcaucus region.

In 1967 Hushang Taraz of the Geological Survey of Iran discovered

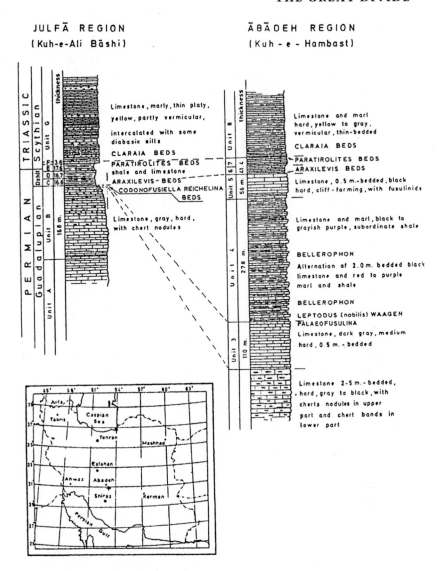

Figure 2.6. Stratigraphic sections and correlation between Permo-Triassic boundary sections at Kuh-e-Ali Bashi (Iranian-Armenian border) and Kuh-e-Hambast (central Iran). From Taraz (1971); reprinted by permission.

59

a very thick Permo-Triassic boundary sequence in the Abadeh region of central Iran, about 1,000 km southeast of Dzhulfa (Taraz 1969, 1971, 1973). The Permian rocks form a transgressive sequence and are divided into the Surmaq, Abadeh, and Hambast formations. They are overlain by Lower and Middle Triassic units (the Limestone Group and the Dolomite Group). The entire sequence is lithologically quite similar to the Dzhulfa sections (figure 2.6). The Abadeh Formation is a thick (400 m) series of black shales capped by a limestone member. Overlying the Abadeh Formation units 6 and 7 (based on their original field designations) comprise the Hambast Formation. Both units 6 and 7 are 17–18-m-thick limestones and shales deposited in shallow marine waters and the upper part of unit 7 contains a 4-m limestone bed with the ammonoids *Paratirolites* but few other fossils. The diversity and abundance of the faunas decreases through units 6 and 7, with marked drops at the boundary between 6 and 7, within unit 7, and at its top. The abundance of boron and lithium within unit 7 and at the boundary of unit 7 and the overlying unit provide evidence of a fresh-water influx into the basin. This and other evidence suggests the peak of a regressive cycle at the top of unit 7 (Iranian-Japanese Research Group 1981), although Taraz (1971) interpreted the facies change as a product of a sudden climatic shift.

Above the *Paratirolites* limestone is a 10–30-cm-thick brownish or greenish shale that is taken as the base of the Triassic. This shale grades up into a stromatolitic limestone indicative of shallow subtidal to intertidal conditions. The only fossils within this unit are algae, ostracodes, ammonoids, bivalves, including *Claraia*, and conodonts. The stromatolitic beds contain the conodonts *Hindeodus typicalis* and *Isarcicella? parva* with *Isarcicella isarcica* appearing in the overlying limestone. Most authors consider the uppermost Permian (Changxingian) and the earliest Griesbachian to be missing from this region (Iranian-Japanese Research Group 1981; Teichert 1990), but the correlation between the Dorashamian Stage and the type Changxingian is uncertain (see below). The ammonoids indicate a late Griesbachian age for the earliest Triassic, while the presence of *Isarcicella? parva* suggests the early Griesbachian is present but the *Otoceras woodwardi* zone is absent (Golshani, Partozar, and Seyed-Emami 1988). The lithologic and faunal similarities between these formations and the sections in the Julfa region are striking and indicate deposition in the same system (Taraz 1969, 1971, 1973; Iranian-Japanese Research Group 1981; Bando 1979; Golshani 1988 provides an update on recent activity).

Pakistan: The Salt Range

Investigation of Permian and Triassic deposits in this region began in 1848 (see Kummel and Teichert 1966, 1970a, the sources for the following discussion). A. B. Wynne examined the area in the late 1800s and the similarity between the Permian and Triassic beds is emphasized by his description of the sequence:

> These Triassic rocks, though lithologically distinguishable, present no such marked contrast to the carboniferous [Permian] formation as exists between the Trias and succeeding beds. The limestones are more thin-bedded, and the shales or marls of different character and somewhat different color; but the whole aspect of the group is such that, were it not for the paleontological evidence, it might pass for a portion of the paleozoic rocks below with which it was classed by Dr. Fleming. (Wynne 1878:97–98; cited in Kummel and Teichert 1970a:8)

W. Waagen visited the area with Wynne and published numerous monographs on the fauna from 1879 to 1895. He recognized a more pronounced break in sedimentation than Wynne and accurately described the magnitude of the faunal discontinuity between the Permian and Triassic. Following his visit to the Salt Range in 1952, Otto Schindewolf concluded the boundary between the Permian and Triassic was continuous, a claim he used to support his theory that the sudden faunal turnover required a catastrophic, extraterrestrial cause, in particular cosmic radiation (see chapter 9). Kummel and Teichert continued their cooperative investigation of Permo-Triassic boundary sections with work in the Salt Ranges during the 1960s.

The latest Permian Chhidru Formation and the earliest Triassic Kathwai Member of the Mianwali Formation are the most important units in the sequence (figure 2.7). The Chhidru contains a highly fossiliferous fauna in sandstone and limestone beds, capped by a 3-m thick fine-grained sandstone known as the White Sandstone Unit. The increasing clastics and the boron content indicates a nearshore depositional environment with increasing fresh-water influx toward the top of the unit. The fossil content of the White Sandstone varies greatly between localities from absent to rich, but at some localities diverse faunas are found up to the very top of the unit (Kummel and Teichert 1970a). Brachiopods and gastropods are common throughout the formation (Pakistani-Japanese Research Group 1985). While the age of the Chhidru Formation has long been a source of controversy,

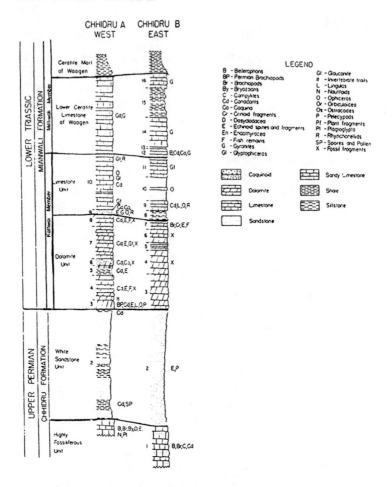

Figure 2.7. Stratigraphic section across the Permo-Triassic section at two localities in the Salt Range, Pakistan. From Kummel and Teichert (1970b).

Furnish and Glenister (1970) failed to find any characteristic Changxingian ammonoids and assigned a Djulfian age to the unit. Grant's (1970) analysis of the brachiopods suggests a longer hiatus may have occurred. On balance, Kummel and Teichert (1973) and Teichert (1990) suggested a gap of the Changxingian stage (including the Dorashamian), and perhaps some of the lowest Griesbachian, a judgment corroborated by the Pakistani-Japanese Research Group (1985).

The base of the Kathwai Member is a thin-bedded to massive dolomite bed that is overlain by a limestone unit (the Pakistani-Japanese Research Group [1985] divided the Kathwai into three units, but I will retain the divisions of Kummel and Teichert [1970b]). The fauna of the two units are identical and variations in thickness of the two units suggest the dolomite is simply a diagenetic overprint on the limestone. Over 28 genera of invertebrates and 3 fish genera have been described from the fauna, including a bryozoan, Permian-type brachiopods (to which Grant [1970] assigned a Djulfian age), ammonoids, nautiloids, gastropods, bivalves, fish, and other fossils. Numerous trails and burrows occur on the surfaces of several beds, including the presumptive Permo-Triassic contact. Kummel and Teichert (1970a) report the characteristic ammonoids *Ophiceras connectens* and *Glyptophiceras himalayanum* and assign a late Griesbachian age to this bed. Conodonts include *Isarcicella? parva* and *Isarcicella isarcica* (Pakistani-Japanese Research Group 1985). The lack of evidence of reworking of the brachiopods, bivalves, or conodonts indicates this mixed faunal assemblage is a transitional Permo-Triassic fauna, much as occurs elsewhere in the world. Kummel and Teichert (1970b) believe the Permo-Triassic boundary is a paraconformity. The Pakistani-Japanese Research Group (1985) disagreed with this interpretation, however, placing the Permo-Triassic boundary between their Lower and Middle units of the Kathwai Member of the Mianwali Formation. The primary references on the Salt Range are Schindewolf (1954), Kummel and Teichert (1966, 1970a), which includes a detailed history of previous research, and Pakistani-Japanese Research Group (1985), and brief reviews are provided by Kummel and Teichert (1973) and Teichert (1990).

Kashmir

As if to prove my contention that most of the Permo-Triassic boundary sections are carefully located beneath civil wars or in otherwise inaccessible regions, one of the best occurs in the disputed region of Kashmir at Guryul Ravine and at a spur near Barus. As described earlier in this chapter, the Permo-Triassic boundary appears to lie within the Khunamuh Formation. The underlying Zewan Formation at Guryul Ravine in Kashmir is lithologically quite similar to the Chhidru Formation, but contains few diagnostic fossils. It is divided into four members (A to D), with the amount of clastic material increasing through the section from limestone and sandy shale to sandy limestones. The alternating limestones and black shales of the overlying Khunamuh Formation have been divided into six members (E to J), based on increasing amounts of limestone. The change in facies suggests an initial deepening of conditions during the Zewan followed by shallowing toward the Zewan-Khunamuh boundary, where an abrupt lithologic boundary occurs, then a return to open marine conditions during the deposition of the Khunamuh Formation (Nakazawa and Kapoor 1981). The lowest member of the Khunamuh (E_1) contains many species recorded from the underlying Zewan Formation, but Nakazawa and Kapoor (1981) record four species persisting from unit E_1 to E_2—*Pustula?* sp., *Marginifera himalayensis*, and the endemic bivalves *Claraia bioni* and *Etheripecten haydeni*, as well as several conodonts. The mixture of Permian brachiopods like *Spinomarginifera* with such Triassic forms as *Claraia bioni* within bed E_1 suggests a gradual faunal transition between the Permian and Triassic (Teichert, Kummel, and Kapoor 1970). The characteristic earliest Triassic ammonoid *Otoceras woodwardi* does not appear until unit E_2, leading to suggestions that *Claraia bioni* in unit E_1 represents the first, latest Permian, appearance of the genus. Ammonoids assigned to the genus *Ophiceras* appear in unit E_2 but first become diverse in unit E_3 (Nakazawa, Bando, and Matsuda 1980). Conodont data may be the most informative, however. Unit E_2 can be divided into two conodont zones, a lower *Hindeodus typicalis* zone, and an upper *Isarcicella? parva* zone that extends into the base of unit E_3. The conodont *Isarcicella isarcica* occurs only in a narrow zone at the base of unit E_3 (Nakazawa, Bando, and Matsuda 1980).

Teichert, Kummel, and Kapoor (1970) place the boundary at the lithologic break between the Zewan and Khunamuh formations (see also Teichert 1990). Nakazawa et al. (1975) place the boundary within unit E_1, while Nakazawa (1981) places the boundary between units E_1 and E_2. The conodont data presented by Nakazawa, Bando, and Matsuda (1980) support a boundary between units E_2 and E_3 (see also Sweet 1992). Obviously the latter position places both *Claraia* and *Otoceras woodwardi* in the latest Permian rather than as diagnostic fossils for the earliest Triassic. Probably the most accurate conclusion is Teichert's (1990:211): "Unit E_1 has, indeed, a fauna of puzzling composition. Nothing like it seems to be known from any other locality, where the oldest Griesbachian sediments are recorded." The primary monographs on the stratigraphy and fauna are Nakazawa et al. (1975), Nakazawa and Kapoor (1981), and Nakazawa, Bando, and Matsuda (1980). Other discussions of these sections include Kapoor and Tokuoka (1985) and Kapoor (1992).

Himalayas

Further to the east earliest Triassic sediments with *Otoceras wood-wardi* are widespread throughout the Himalayan region but no continuous boundary sequences are known, probably because the region was emergent during latest Permian time before being inundated during the earliest Triassic (Kapoor and Tokuoka 1985). A diverse brachiopod fauna from the Marsyangdi Formation near Manang in central Nepal has been reported (Waterhouse and Shi 1990). The 165 brachiopod species, mostly productids, make it the most diverse Late Permian brachiopod assemblage yet recorded. The overlying Pengba Member of the Kangshar Formation contains *Otoceras*, *Claraia*, and other early Triassic fossils but is separated from the Marsyangdi Formation by a disconformity. The length of this hiatus is unknown, particularly since there is no age control from conodonts on the Marsyangdi Formation.

South China

Kummel and Teichert recognized the significance of Permian sections in South China during the 1960s, but the knowledge gained by Chinese geologists over the past 20 years has increased the importance of the region far beyond what Kummel and Teichert suspected. Latest Permian rocks outcrop across thousands of square kilometers (figure 2.8) and the number of Permo-Triassic boundary sections probably surpasses the rest of the world combined; Chinese geologists have analyzed over 40 sections in detail (Yang et al. 1987; Yang and Li 1992).

South China lay in eastern Tethys during the Permian, although the precise geographic position of the block remains unclear (see chapter 5). Sediments were deposited in an epicontinental sea divided by uplands into a series of platforms and basins (Sheng, Rui, and Chen 1985) and episodic volcanism was common. The Upper Permian sequence has been divided into two regional stages, the lower Wujiaping Stage and the upper Changxing Stage. Within the Wujiaping, the Wujiaping Formation is dominated by carbonates and is rich in foraminifera but poor in ammonoids. In contrast the Longtan Formation is a clastic-rich facies with abundant ammonoids. The Changxing

65

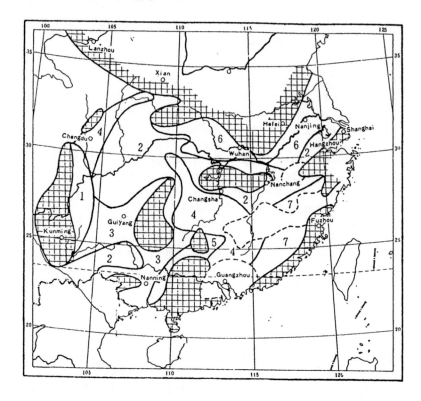

Figure 2.8. Distribution of latest Permian rocks in South China. Cross-hatched areas are exposed regions. From Sheng, Rui, and Chen (1985).

Formation and its lateral equivalents, including the Dalong Formation, are also distinguished by facies shifts (but see Yang et al. 1987). The fauna of the Changxing is far more diverse than that of the underlying Wujiapingian, reflecting the development of shelfal and basinal environments (Li et al. 1991). The Permo-Triassic boundary in South China is seemingly coincident with a change in lithology as the gray biomicrites of the Changxing give way to the marls of the Lower Triassic Yinkeng Formation. The lower member of the Changxing, the Baoqing Member, is a 23-m-thick series of dark gray bituminous limestone with occasional clay layers, some of which may be of volcanic origin. The overlying Meishan Member is about 18 m thick and consists of dark bedded limestones with intercalated chert and clay layers. Permian rocks are overlain by an earliest Triassic unit endowed with a variety of names in different provinces, including the Chinglung, Feixianguan, Daye, and Yinkeng formations. The difference in formation names reflects regional variations in depositional environments across South China (Zhao et al. 1981; Sheng et al. 1984; Sheng, Rui, and Chen 1985; Yang and Yin 1987; Yang et al. 1987).

The Meishan coalfield near Changxing, Zhejiang Province, and southeast of Nanjing, is one of the most important sections (figure 2.9). Furnish and Glenister (1970) described the Changxingian Stage

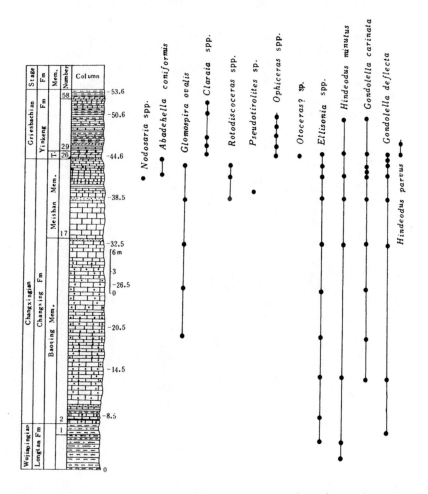

Figure 2.9. Stratigraphic section of the Permo-Triassic boundary at the Meishan coal field, South China. From Yin et al. (1988).

as the youngest global Permian Stage on the basis of sections at Meishan. Here the Changxing sediments were deposited in shallow intertidal to lagoonal settings on a platform slope (Yang and Jiang 1981; Yang and Li 1992). Through the sequence at Meishan marine invertebrates suffer a 53 percent extinction during the Wujiaping Formation, a 32 percent drop at the transition from the Baoqing Member to the Meishan Member, and finally a 91 percent extinction rate during the last section of the Changxing, where 435 of 476 species become extinct (figure 2.10; Yin, Xu, and Ding 1984, cited in Xu et al. 1989). Among ammonoids, the 98 percent species level extinction coincided with the disappearance of the siliceous deposits in which they are found. Bivalves suffer an 85 percent extinction at the generic level; scallops are particularly affected, while euryhaline and infaunal bivalves appear to have been least affected. In South China fusuline foraminifera, largely confined to shallow water, declined by 75 percent at the base of the Wujiaping Stage and only 6 genera survived to the end of the Changxingian Stage; none is found in the earliest Triassic. In contrast, a total of 30 genera of nonfusuline forams have been recovered from the Changxingian Stage of which 11 survived into the

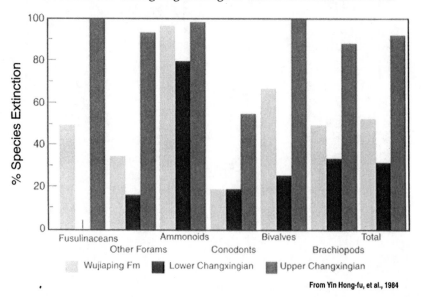

From Yin Hong-fu, et al., 1984

Figure 2.10. Species extinction rates for the most important groups in the Wujiaping, Lower Changxing, and Upper Changxing formations in South China. From Yin, Xu, and Ding (1984), translated in Xu et al. (1989).

Triassic. Most of these were living in water depths greater than 20 m, which may have contributed to their survival.

A transitional bed of clays, silts, and limestones varying in thickness from 15 to 50 cm separates obvious Permian and Triassic sediments and contains a remarkable mixed faunal assemblage, one which is far more developed than mixed faunal assemblages elsewhere (figure 2.11). This assemblage was identified in 21 of the 31 sections studied by Sheng et al. (1984) and has been subsequently recognized at additional localities. In South China the mixed fauna contains Triassic-type ammonoids and bivalves and Permian-type brachiopods. Almost all of the brachiopods are derived from those found in the siliceous facies of the Meishan Member (Liao 1980). At Meishan these transitional beds are 26 cm thick, typical for most of the marine transitional beds, and three distinct mixed-faunal assemblages have been recognized. The lowest unit, mixed bed 1, is a yellowish to greenish shale about 10 cm thick capping a thin clay layer. Uncertainty about the age of this clay was resolved by the discovery of a variety of marine fossils, including the characteristic latest Permian conodonts *Neogondolella changxingensis*, *N. deflecta*, and *Hindodella minutus* (Rui et al. 1988). Mixed bed 2 is a 20-to-30-cm-thick marly dolomite, and mixed fauna bed 3 is largely greenish shale with limestone beds (Sheng et al. 1984).

The transitional assemblages at Meishan represent a widespread carbonate mud assemblage (Yin 1985a). The faunal affinities of the brachiopods, bivalves, cephalopods, and forams (which comprise the bulk of the fauna) were cataloged by Sheng et al. (1984) (figure 2.12m). In mixed bed 1, 19 of the 34 species (56 percent) were of Permian affinities, all 14 species recovered from mixed bed 2 were from the Permian, but only 2 of 13 species (15 percent) in mixed bed 3. Of the 56 genera and 90 species recorded, 13 genera are bivalves, including several species of *Claraia*, 11 are brachiopods, 8 are ammonoids, including *Otoceras*, 2 are gastropods, 8 are conodonts, including *Isarcicella? parva*, 3 are foraminifera, 5 are plants, and 1 insect genus was recovered (Yang et al. 1987; Yin 1985a). Tozer (1979) argued that the Permian fossils within these mixed faunal assemblages had been reworked from older sediments, but subsequent studies have confirmed that the assemblages are real, not a taphonomic artifact (e.g., Bottjer, Droser, and Wang 1988). Mixed faunal assemblages have also been reported from the Dalong Formation in western Zhejiang Province (Fu 1988).

The second type of transitional assemblages is more restricted,

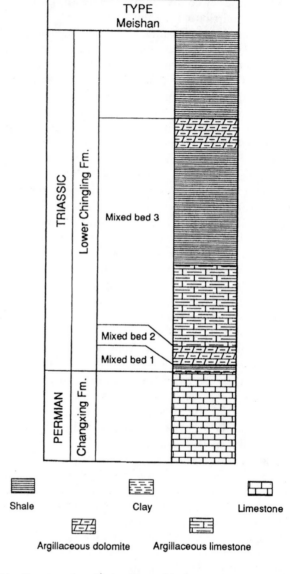

Figure 2.11. Stratigraphy of mixed faunal beds.

Horizon / Species	Mixed 1 P.*	T.*	Mixed 2 P.	T.	Mixed 3 P.	T.	Above Mix.3 P.	T.
BRACHIOPODS								
1. *Lingula* sp.	1●							
2. *Acosarina* sp.	2●							
3. *Fusichonetes pigmaea* (LIAO)	3●		3●					
4. *F. nayongensis* (LIAO)	4●							
5. *F.* sp.	5●							
6. *Waagenites barusiensis* (DAVIDSON)	6●		6●					
7. *W.* cf. *barusiensis* (DAVIDSON)	7●							
8. *W.* sp.	8●		8●					
9. *Paryphella sulcatifera* (LIAO)	9●							
10. *P. orbicularis* (LIAO)	10●		10●					
11. *P.* sp. nov.	11●							
12. *P. triquetra* LIAO	12●		12●					
13. *P.* sp.	13●		13●					
14. *Neowellerella pseudoutah* (HUANG)	14●		14●					
15. *N.* sp.					15●		15●	
16. *Crurithyris flabelliformis* LIAO			16●					
17. *C. subspeciosa* LIAO			17●					
18. *C.* sp.			18●		18●			
19. *Paracrurithyris pigmaea* (LIAO)	19●		19●					
20. *P.* sp.			20●					
21. *Araxathyris araxensis* GRUNT	21●							
22. *A.* cf. *araxensis* GRUNT	22●							
23. *A. minuta* GRUNT	23●							
CEPHALOPODS								
24. *Grypoceras* sp.		24●						
25. *Pseudogastrioceras* sp.	25●							
26. *Pseudosageceras* sp.		26●						
27. *Otoceras* sp.		27●						
28. *Tompophiceras* sp.		28●						
29. *Metophiceras* sp.		29●						
30. *Hypophiceras* cf. *martini* TRÜNPY		30●						
31. *H.* sp. nov.		31●						
32. *H.* sp.		32●						
33. Ophiceratid		33●				33●		
34. *Ophiceras* sp.						34●		
35. *Lytophiceras* sp.						35●		
BIVALVES								
36. *Claraia wangi* (PATTE)						36●		
37. *Cl. longyensis* CHEN						37●		
38. *Cl. griesbachi* BITTNER						38●		
39. *Cl.* cf. *painkhandana* (BITTNER)						39●		
40. *Cl. dieneri* NAKAZAWA						40●		
41. *Cl. fukienensis* CHEN						41●		
42. *Cl.* ? sp.		42●						
43. *Cl. stachei* (BITTNER)						43●		
44. *Peribositra baoqingensis* CHEN		44●						
45. *Entolium* sp.		45●						
46. *Bakevellia* sp.		46▲						
47. *Eumorphotis* ? sp.		47●						
48. *Pteria ussurica variabilis* CHEN et LAN		48●						
49. *Palaeonucula* sp.						49●		
FORAMINIF.								
50. *Nodosaria* sp.			50▲					
51. *Pseudoglandulina* sp.			51●					

Figure 2.12. Composition of the three mixed faunal assemblages and the affinities, either Permian or Triassic, of each species. From Sheng et al. (1984).

both geographically and in terms of faunal diversity. Beds of this near-shore, clastic-dominated transitional sequence are much thicker than those of the carbonate facies and contain a restricted fauna. Erosional contacts occur in some of these sections but are uncommon in the marine transitional sequences.

The boundary clay in the Meishan section is clearly of volcanic origin. Indeed, evidence of explosive silica-rich volcanism is widespread in South China. This volcanic ash has been altered to a montmorillonite or montmorillonite-illite clayrock, and commonly contains tuffaceous texture, bipyramidal quartz, spherules, volcanic shards, and high-temperature quartz (Yin et al. 1989, 1992). In some regions the volcanic ash is relatively unaltered and at many localities several clay layers occur throughout the boundary sequence. For example, at Meishan volcanic clays are found at the base of mixed fauna bed 1, mixed fauna bed 3, and again later during the deposition of mixed fauna bed 3 (He et al. 1987). In Guizhou Province an Upper Permian sequence includes at least ten volcanic clay layers (Yin et al. 1989). Yin et al. (1992) estimate the boundary clay at the base of mixed fauna bed 1 covers more than 1 million square kilometers across 12 provinces in southern China. Trace element analysis of boundary sections at Meishan, Shangsi, and Liangfenya indicates the clays are enriched in cesium, zirconium, hafnium, tantalum, and thorium, and depleted in chromium, cobalt, and iridium, consistent with eruption from a massive, silica-rich volcanic source (Zhou and Kyte 1988). It is important to note that not all boundary clays are altered volcanic ash; some contain purely terrigenous material. Not surprisingly, a number of paleontologists have implicated these massive volcanic events as a cause of the extinction, a topic to which we will return in chapter 9.

Reports of anomalously high concentrations of iridium within the boundary clays (Sun et al. 1984; Xu et al. 1985) have not been substantiated (Orth 1989; Clark, Wang, and Orth 1986; Zhou and Kyte 1988), although some Chinese paleontologists continue to press their claim (Xu et al. 1989). Obviously confirmation of iridium enrichment would support an extraterrestrial impact, and the evidence for and against this suggestion will also be carefully considered in chapter 9.

Among the youngest Paleozoic reefs are those found within the Changxing Formation in Sichuan and Hubei Provinces of southwestern China. The earliest phases of reef development at these localities involves stabilization of carbonate shoals by brachiopods and crinoids, then the appearance of bryozoans. As water depth increases a diverse fauna of brachiopods, bryozoans, crinoids, and foraminifera develops, followed by the establishment of reefs dominated by 12 or more genera of calcareous sponges along with various encrusting organisms including bryozoans, foraminifera, and polychaete worms, occasional rugose corals, brachiopods, echinoderms, and molluscs. The reefs are overlain by lagoonal sediments and the Permo-Triassic boundary is marked by brief subaerial exposure. Reefs of this complexity in the very latest Permian certainly belie claims for gradual environmental deterioration, at least in South China, until after these reefs disappear. A 1–6-cm yellowish-gray clay layer apparently corresponds to the boundary clays elsewhere in South China (Reinhardt 1988a, b). The disappearance of most taxa is associated with the elimination of the reefal environment and the appearance of an evaporitic intersupratidal facies, not a sudden extinction at the boundary horizon (Reinhardt 1988b).

The literature on the South China sections is extensive and continues to grow rapidly. Although many papers are published only in Chinese, a growing number are either translated or contain a lengthy synopsis in English. Critical references include Zhao et al. (1981), Li et al. (1986, 1989, 1991), Sheng et al. (1984, 1987), Sheng, Rui, and Chen (1985), and Yang et al. (1987).

IMPLICATIONS FOR THE EXTINCTION

THIS OVERVIEW of Permo-Triassic boundary sections has purposely left several related issues unresolved, particularly the nature of the transitional mixed-faunal assemblage, the question of the continuity of the boundary sections, and the correlations between the various sections.

The Mixed Faunas

Yin (1985b) divided these mixed-faunal assemblages into three classes, reflecting the complex nature of biotic transition between the Changxingian and Griesbachian. The first class includes beds like those in East Greenland that are clearly a result of a reworking of Permian fossils into early Triassic sediments. The second class includes sequences in which the mixed assemblage is real but overlies a condensed interval during the latest Permian. This includes sections in the Salt Range, the Himalayas, and the Dinwoody Formation in the western United States. The mixed-faunal assemblages in South China and the southern Alps appear to occur within biostratigraphically continuous sequences, and Yin would restrict use of the term *transitional bed* to these units.

Unfortunately the recent growth in the discipline of taphonomy has yet to touch the study of the Permo-Triassic boundary. These mixed-faunal assemblages are clearly in need of detailed taphonomic investigation to resolve the nature of their formation. For example, the dramatic stable isotope change during the latest Permian (see chapter 7) permits isotopic analysis of individual fossils. Reworked specimens will have signatures of the level at which they were originally deposited.

Continuity of Marine Boundary Sections

The continuity of deposition across the Permo-Triassic boundary continues to be a highly contentious issue in nearly all of the regions described above. Since the global boundary stratotype should be placed at a reasonably continuous section, stratigraphers are particularly interested in this issue and there is considerable incentive for champions of particular sections to denigrate competing sections.

In South China the Permo-Triassic contact was originally described as unconformable, but intensive study of the sections convinced Zhao et al. (1981) that the sections were generally conformable and without a major hiatus at the Permo-Triassic boundary. Most Chinese researchers have held similar views (e.g., Sheng et al. 1984; Yang and Yin 1987; Yang et al. 1987). In contrast, several western geologists have found evidence of discontinuities across the presumed Permo-Triassic boundary in China (e.g., Teichert 1990; Tozer 1979).

Virtually all discussions of sedimentologic continuity across the boundary proceed from the vacuous assumption that sedimentation may in principle be continuous, thus confusing the issue of whether sequences are conformable (whether a demonstrable hiatus occurs at the boundary) with continuity. As Peter Sadler demonstrated in 1981, *all* geologic sections are replete with gaps in sedimentation of varying length. The question of continuity (or completeness in Sadler's terms) is meaningless, absent discussion of the time scale under observation. In other words, if the question is whether or not some sediment was deposited during a 100,000-year interval, a gap of 10,000 years, or even 99,000 years, is acceptable since some sediment was deposited during that 100,000-year span. On the other hand, if we are interested in examining individual 5,000-year intervals (the level of resolution) within that 100,000 years, gaps are more likely and the completeness of the section decreases. Thus without a discussion of the level of resolution there is little point in discussing completeness.

Returning to the Permo-Triassic boundary, if the sections are conformable (no obvious erosional breaks), the question of completeness is limited by the ability to resolve time. Here the maximum achievable resolution is a single conodont zone (on average, probably one million years). If all conodont zones are present the section is complete at the level of time represented by a conodont zone. There are almost certainly gaps of less duration and Sadler's techniques provide one way of characterizing them, although doing so provides little additional information about the extinction. One final point: sedimentation rates certainly vary between sections and condensed sections (where the sedimentation rates are low for a lengthy interval) may add further complexity—they combine everything together. For example, Magaritz (1992) has argued that the gradual change in stable isotopes in the southern Alps relative to the rapid shift in South China (chapter 7) indicates that the Alpine sections are more complete. A more likely explanation is simply differences in sedimentation rate (a slower rate in South China) and not more or larger gaps in South China.

Correlations and Correlation Problems

The biostratigraphic information from these sections has been used to define a series of biostratigraphic zones for different groups, each defined on the basis of the distribution of characteristic genera or species (figure 2.13). For example, the *Otoceras* ammonoid zone defines an interval of time when *Otoceras* was very abundant. In some cases the genus or species that defines the interval occurs only during that interval; in other cases, the taxon simply reaches peak diversity. Although a variety of biostratigraphic zones are included in figure 2.13, ammonoids and conodonts have proved to be the most useful. A glance at the chart illustrates why. The Changxingian contains a single bivalve zone but two conodont zones, and the finer zonation of the conodonts makes them a more useful tool in biostratigraphy. Brachiopods can also be divided into two zones during the Changxingian but are generally found less frequently than conodonts and are thus less useful. The attractiveness of conodonts for biostratigraphy is illustrated by the presence of three different zonal schemes in figure 2.13. These reflect differences of opinion among conodont specialists on correlation between stratigraphic sections and thus on the times of first and last appearance of different conodont species.

These correlation problems are at the heart of many of the disputes over the nature of the Permo-Triassic boundary. Various correlations for the Permian, particularly the Late Permian, have appeared over the years (Anderson 1981; Waterhouse 1978), but I have relied upon

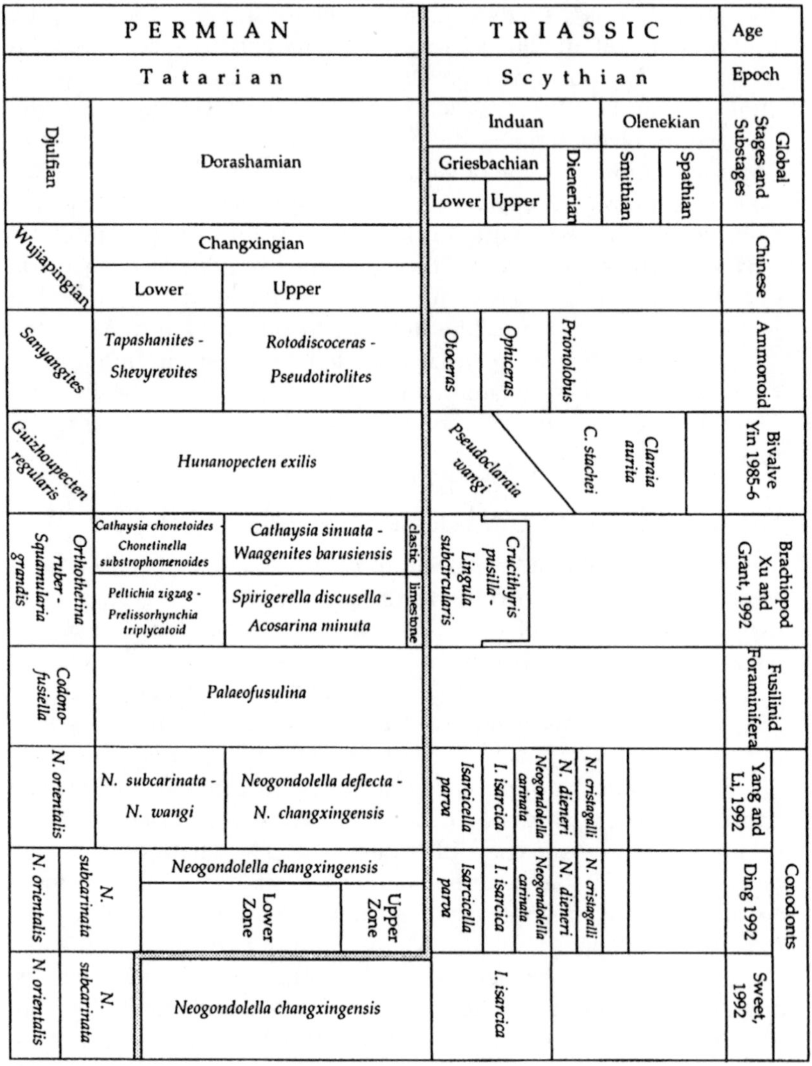

Figure 2.13 — Biostratigraphic zonation chart (Late Permian–Early Triassic).

PERMIAN			TRIASSIC				Age
Tatarian			Scythian				Epoch

Global Stages and Substages (Triassic portion):

	Induan		Olenekian		
Djulfian / Dorashamian	Griesbachian	Dienerian	Smithian	Spathian	
	Lower	Upper			

Chinese

Wujiapingian / Changxingian		Triassic (blank)
Lower	Upper	

Ammonoid

Sanyangites			Otoceras	Ophiceras	Prionolobus
Tapashanites - Shevyrevites	Rotodiscoceras - Pseudotirolites				

Bivalve (Yin 1985-6)

Guizhoupecten regularis / Hunanopecten exilis	Pseudoclaraia wangi	C. stachei	Claraia aurita

Brachiopod (Xu and Grant, 1992)

Orthothetina ruber - Squamularia grandis	Cathaysia chonetoides - Chonetinella substrophomenoides	Cathaysia sinuata - Waagenites barusiensis	clastic / limestone	Crucithyris pusilla - Lingula subcircularis
	Peltichia zigzag - Prelissorhynchia triplycatoid	Spirigerella discusella - Acosarina minuta		

Fusilinid Foraminifera

Codono-fusiella / Palaeofusulina

Conodonts (Yang and Li, 1992)

N. orientalis	N. subcarinata - N. wangi	Neogondolella deflecta - N. changxingensis	Isarcicella parva	I. isarcica	Neogondolella carinata	N. dieneri	N. cristagalli

Conodonts (Ding 1992)

N. orientalis	N. subcarinata	Neogondolella changxingensis		Isarcicella parva	I. isarcica	Neogondolella carinata	N. dieneri	N. cristagalli
		Lower Zone	Upper Zone					

Conodonts (Sweet, 1992)

N. orientalis	N. subcarinata	Neogondolella changxingensis	I. isarcica

Figure 2.13. Biostratigraphic zonation for the Late Permian-Early Triassic. Based on a variety of sources, as discussed in the text.

recent work, particularly papers in Sweet et al. (1992), to prepare a chart showing the correlations between the sections described above (figure 2.14), based on the work of a great many stratigraphers.

Those interested in the basis of this correlation and the problems in correlating between sections should consult papers in Cassinis (1988), Logan and Hills (1973), and Sweet et al. (1992), as well as Iranian-Japanese Research Group (1981), Kozur (1989), Sheng et al. (1984), Teichert (1990), Tozer (1979, 1988a, b), Yang and Yin (1987), Yin (1985b), and Yang et al. (1987). I will limit my discussion to the related problems of the relationship between the Dorashamian and Changxingian Stages, and between these stages and the overlying *Otoceras woodwardi* zone. The boundaries between the Dorashamian and Changxingian sequences remain problematic. Schenck et al. (1941) defined the Dzulfian Stage for uppermost Permian rocks in the Armenian region containing a distinctive faunal assemblage with the ammonoid *Prototoceras*. The upper boundary of the Dzulfian was presumably the appearance of the *Otoceras*-bearing beds of the Triassic. Subsequently, as noted above, Ruzhentsev and Sarycheva (1965) considered the *Paratirolites* limestone to be lowest Triassic but Rostovtsev and Azaryan (1973) rejected this position, placing these units in their new Dorashamian Stage. The use of the term *Dorashamian* has varied considerably between different geologists, but it is characterized by a distinctive assemblage of ammonoids, fusulinids, and conodonts (Iranian-Japanese Research Group 1981).

The Changxingian Stage was proposed by Furnish and Glenister (1970) as the latest Permian Stage based on sequences in South China. Since the base of the Changxingian and the Dorashamian are both defined by the extinction of araxoceratid ammonites and the first appearance of the ammonite *Shevyrevites*, their lower boundaries are isochronous. The difficulty lies in the relationship between the upper boundaries. Zhao et al. (1981) and Yang and Yin (1987) maintained that the Dorashamian is equivalent to the lower part of the Changxingian, while Rostovtsev and Azaryan (1973) and Kozur (1989) consider the Changxingian to be correlated to the lower Dorashamian. Other authors have maintained that the two are basically equivalent (e.g., Sheng et al. 1984). The question remains open but the preponderance of evidence appears to support the equivalence of the two stages.

Choosing between these alternatives involves the troubling assumption that the *Otoceras woodwardi* zone lies directly on top of the *Paratirolites/Pseudotirolites* zone of the Dorashamian/Changxin-

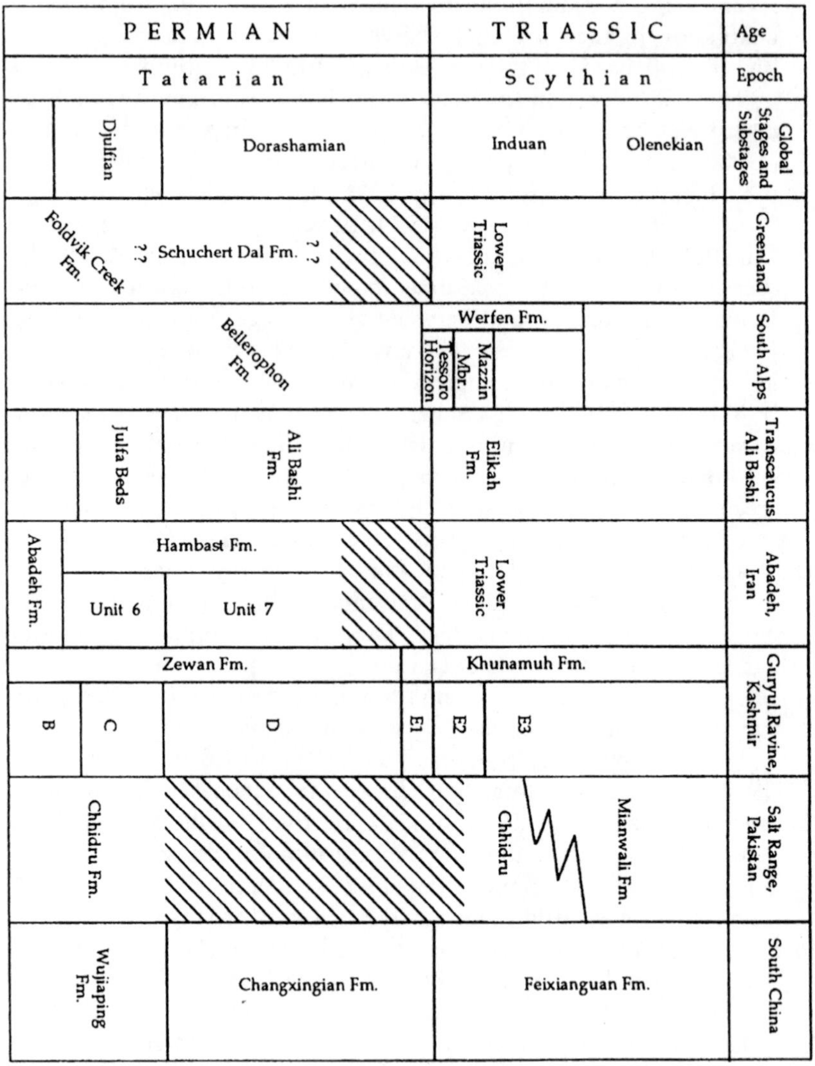

PERMIAN			TRIASSIC		Age
Tatarian			Scythian		Epoch
Djulfian	Dorashamian		Induan	Olenekian	Global Stages and Substages
Foldvik Creek Fm.	?? Schuchert Dal Fm. ??	/////	Lower Triassic		Greenland
Bellerophon Fm.		Werfen Fm. / Tessoro Horizon / Mazzin Mbr.			South Alps
Julfa Beds	Ali Bashi Fm.		Elikah Fm.		Transcaucus Ali Bashi
Abadeh Fm.	Hambast Fm.	/////	Lower Triassic		Abadeh, Iran
	Unit 6	Unit 7			
Zewan Fm.			Khunamuh Fm.		Guryul Ravine, Kashmir
B	C	D	E1	E2	E3		
Chhidru Fm.	/////		Chhidru	Mianwali Fm.	Salt Range, Pakistan
Wujiaping Fm.	Changxingian Fm.		Feixianguan Fm.		South China

Figure 2.14. Stratigraphic zonation for major Permo-Triassic boundary sections.

gian stages (Sweet et al. 1992). The lack of evidence to support this assumption led to suggestions that the Chinese Changxingian Stage may be in part equivalent to the *Otoceras* zone in the Himalayas (Yin et al. 1988; Sweet 1992). This now seems to be correct, suggesting that either some section is missing in the Himalayas or the extinction peak in South China occurred substantially later than in western Tethys. Hence the question is not without interest.

Otoceras is largely confined to temperate latitudes while *Paratirolites/Pseudotirolites* has a tropical distribution (Yin et al. 1988). The discovery of fragments of *Otoceras* sp. at Changxing that overlie the *Pseudotirolites* Zone mark the spot where the two genera occur in the same locality (Tozer 1979; Sheng et al. 1984). Numerous authors have commented on the appearance of *Otoceras* in beds otherwise of Permian aspect, and Waterhouse suggested that a second *Otoceras* zone, the *O. concavum* zone, underlies *O. woodwardi* in the uppermost Permian (see also Bando 1980). This argument was rejected by Tozer (1979, see also Tozer 1988a), who argued that *Paratirolites vediensis* at Kuh-e-Ali Bashi in Iran are the same species as *Shizoloboceras fusuiense* at the upper Changxingian at Meishan and thus that the upper Dorashamian and upper Changxingian are correlated.

Conodonts may resolve this dispute. Yin et al. (1988) argue that the conodont *Isarcicella? parvus* provides a better basis for defining the earliest Triassic than do ammonites. *Isarcicella?* appears in the transitional beds in South China and graphic correlation between Chinese and Himalayan sections suggests overlap between the *Otoceras* and *Pseudotirolites* zones (Sweet 1979, 1992; Yin et al. 1988). Graphic correlation is a simple quantitative correlation technique that involves choosing a section to serve as the standard reference section, then plotting the first and last appearances of species in that section against a second section. If the data are good and the depositional settings were similar, the first and last appearances should form a straight line with the slope of the line representing the relative rate of sediment accumulation between the sections. Gaps in one of the sections will show up as kinks in the line. A composite section is then constructed by selecting, and plotting, the lowest first occurrence value from the two sections, and the highest last occurrence. A new section is then plotted against this composite section and the procedure repeated with as many sections as available (see Sweet 1992). The result is a correlation of rock units in terms of their

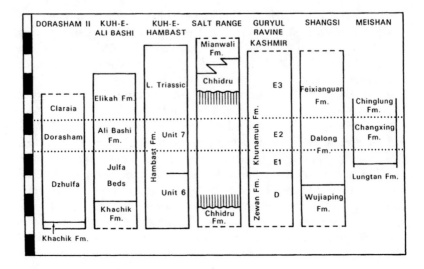

Figure 2.15. Alternative correlation based on graphic correlation of conodonts. From Sweet 1992.

fossils. Sweet's results for seven boundary sections is shown in figure 2.15 and the species range chart in figure 2.16. This correlation supports claims by other workers that the Dorashamian beds and the Ali Bashi Formation are equivalent to unit E_2, the *Otoceras woodwardi* zone of the Khunamuh Formation at Guryul Ravine, Kashmir. Tozer (1988a, b) has strongly criticized these results, largely on the basis of the ammonoid distributions discussed above. This debate continues and remains one of active discussion.

PICKING THE PERMO-TRIASSIC BOUNDARY

THESE DIFFICULTIES in biostratigraphic correlation between sections increase the difficulties in identifying the position of the Permo-Triassic boundary. Recall that by the 1920s the contact between the Zechstein Formation and the overlying Werfen Formation was taken as the boundary and *Otoceras woodwardi* and *Claraia* were employed to identify the earliest Triassic sediments elsewhere in the world. Yet today several different positions for the boundary have been advanced just for the section at Guryul Ravine in Kashmir, as discussed earlier in this chapter. How did this come about? Traditional correlations place the boundary at the base of the *Otoceras* bed, by correlation with the Buntsandstein-Werfen formations in the Alps (level A in figure 2.17) (Kummel 1972, 1973a; Kummel and Teichert 1973; Tozer 1979, 1984, 1988a, b; Nakazawa et al. 1975). A slight modification was proposed by Yin (1985b; see also Yin et al. 1988; Yang and Yin 1987). Yin took the base of the transitional beds in South China as the boundary and the *Isarcicella? parva* conodont zone as the marker for the boundary (marked B in figure 2.17). Of course, Sweet's graphic correlation of conodonts suggests the base of the *Neogondolella changxingensis* zone is correlative with the first appearance of *Otoceras woodwardi* and thus he would place the boundary at level C (figure 2.17).

An alternative is to place the boundary between the *Otoceras* zone and the succeeding, *Ophiceras* ammonoid zone (Li et al., 1989, 1991; Kozur 1989; Kotlyar 1991). In South China this would place the boundary at the top of the transitional beds. Li et al. (1991) claim this level is equivalent to the boundary between the *Isarcicella? parva* and *Isarcicella isarcica* conodont zones but they appear to have miscorrelated the ammonoid and conodont zones.

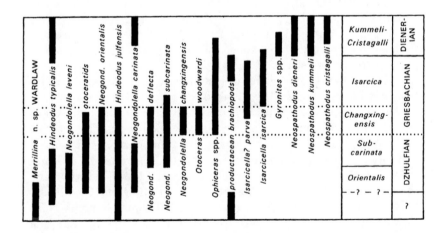

Figure 2.16.   Biostratigraphic ranges of stratigraphically important species based on the graphic correlation technique of Sweet. From Sweet (1992).

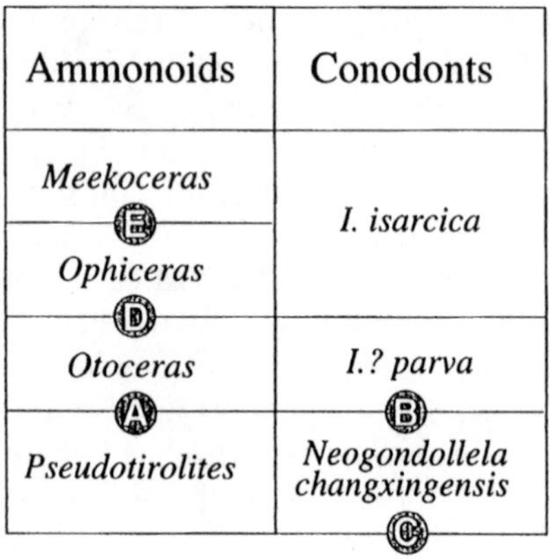

Ammonoids	Conodonts
Meekoceras ⒠	I. isarcica
Ophiceras ⒟	
Otoceras ⒜	I.? parva ⒝
Pseudotirolites	Neogondollela changxingensis ⒞

Figure 2.17. Alternative positions for the Permo-Triassic boundary: A) traditional; B) by Yin (1985); (C) by Sweet (1992) based on an alternative correlation between South China and sections in Western Tethys; (D) by Li et al. (1991) and Kozur (1989); (E) by Newell (1978, 1988). Level C appears to be the most reasonable position, and corresponds with the major number of geochemical and geologic perturbations.

Norman Newell of the American Museum of Natural History in New York and the dean of American Permian paleontologists, has long argued that the Griesbachian should be assigned to the Permian and the boundary placed at the top of this stage (Newell 1978, 1988). This level is marked by the appearance of abundant meekoceratid ammonites (the *Gyronites* zone) and the conodont *Neospathodus*. The end of the Griesbachian also marks the beginning of the recovery phase and the first appearance of widespread, easily correlatable faunas. In addition, a normally polarized paleomagnetic reversal ends at this horizon and the $\delta^{13}C$ again becomes positive. While acknowledging the historical primacy of the *Otoceras* beds as the base of the Triassic, Newell claims that the extent of the Permian radiation of ceratites (the group of ammonites to which *Otoceras* and *Ophiceras* belong) was unknown in the last century, and if Mojsisovics, Diener, and Griesbach had known the distribution of the group they would have placed *Otoceras* and *Ophiceras* in the Permian (a point Tozer [1988a, b] disputes). Moreover, Newell accurately points out that extensive morphologic variation in Late Permian otoceratid ammonites makes it difficult to reliably separate different genera. Thus he argues from a purely practical standpoint that the radiation of the meekocerids is a more easily correlated horizon.

Finally, Waterhouse (1973, 1978) proposed an even higher level, the Dienerian/Smithian boundary within the Early Triassic, arguing that faunas below this are primarily of Permian affinities and that true Triassic faunas, particularly ammonites, do not appear until this level. However, this proposal has attracted little support.

Levels A and B in figure 2.17 are roughly equivalent and correspond to the historical boundary. This horizon is marked by volcanic clays in South China, a global shift in carbon isotopes, the appearance of a mixed fauna, and a paleomagnetic reversal. In the view of many stratigraphers this horizon correlates with either the base of the Tesero Horizon in the lower Werfen Formation, or with a level about one-third of the way up the bed. A growing body of opinion supports this point for the Permo-Triassic boundary. However, continued attention must be given to the possibility of correlation problems raised by Sweet (1979, 1992) and Yin (1985b). If they are correct, the generally accepted boundary is diachronous, the problem that got us into this mess in the first place!

Age of the Boundary

So how old is the boundary? These correlation problems and the absence of igneous or metamorphic rocks amenable to radiometric dating have long made the age of the boundary a contentious issue. In his 1973 review, Banks notes that prior to 1964 an age of 225 million years was widely accepted, but that potassium-argon analyses between 1964 and 1973 favored a date of about 235 million years. More recently the date has fluctuated between 250 and 245 million years, although few well-constrained dates have been available (see Menning 1991). Zircons within the boundary clay in South China have provided the first direct dates on the boundary interval (Claoue-Long et al. 1991). The 5-cm-boundary clay between the Changxing Formation and the overlying mixed fauna bed 1 of the Chinglung Formation at Changxing was sampled. Thirty-five zircons were analyzed using a SHRIMP ion microprobe for $^{206}Pb/^{238}U$. After removal of three anomalous results, the remaining crystals produced a date of 251.2 ± 3.4 million years for the crystallization of the ash.

IN SUMMARY, it should be apparent that while major advances have been made in understanding Permo-Triassic boundary sections, considerable work still remains to be done. As always, the search for new boundary sections continues. The results from restudy of the Bellerophon-Werfen boundary demonstrate the need to periodically reexamine sections that have been thought to contain major gaps. The difficulties in correlating between various regions indicate both the inherent problems this boundary poses and our ongoing requirement for enhanced correlation techniques.

III

The Farewell Symphony

*The way in which many Paleozoic life forms disappeared
towards the end of the Permian Period brings to mind
Joseph Haydn's Farewell Symphony where, during the last
movement, one musician after the other takes his
instrument and leaves the stage until, at the end,
none is left.*
TEICHERT 1990:231

THE TRADITIONAL reason for analyzing differences in patterns of extinction between different groups is simple curiosity or to uncover patterns of taxonomic, ecologic, or biogeographic selectivity that reveal something of the causes of the extinction. In the preceding chapter we looked at extinction patterns within individual sections close to the boundary. In this chapter we step back and look at patterns of extinction and survival within individual groups. The information in this and the following chapters (on terrestrial extinction patterns) will be employed in chapter 9 to test the various hypotheses suggested for the extinction, but we must not lose sight of the opportunities created by mass extinctions and the fact that patterns of extinction also determine the course of the postextinction recovery, the topic of chapter 10.

The directions evolution takes following a mass extinction depend on the nature of the extinction. Paleontologists generally acknowledge three possibilities, which Raup recently labeled the "Field of Bullets," "Fair Game," and "Wanton Destruction" scenarios (Raup 1991; see also Erwin 1990a). A Field of Bullets mass extinction can be visualized as a concentrated hail of bullets that randomly encounter (and destroy) individuals. On probabilistic grounds species with smaller populations will disappear more often than species with large populations and there is little selectivity in the pattern of extinction. If only a few species survive, random chance will favor the survival of some groups relative to others, enhancing the differences between pre- and postextinction faunas (Raup 1991). Darwinian fitness provides an advantage in the Fair Game scenario, in which the best-adapted species survive preferentially. Finally, according to the Wanton Destruction scenario the extinction is selective, but the basis of the selectivity is not increased fitness but some other factor.

Which of these scenarios is most accurate is critical to understanding the effect of mass extinctions on the history of life. If the Hail of Bullets scenario is correct, mass extinctions neither advanced nor retarded the adaptive trends established before the extinction and mass extinctions had no long-term effect on the history of life. The Fair Game scenario actually advances adaptive trends already established, so a mass extinction might actually speed up long-term evolutionary trends. Meanwhile, the Wanton Destruction scenario is the most intriguing, and has received wide attention in the past several years (Jablonski 1986a, 1989). In this case the evolutionary clock is reset by the mass extinction, fundamentally reordering the relationships between different groups and setting evolution on a new course. Determining which of these patterns most accurately describes extinction patterns during the Late Permian is one of the objectives of this chapter. In addition to detailing individual extinction patterns we will also consider aspects of selectivity and the rate of extinction.

GENERAL EXTINCTION PATTERNS

FROM THE Guadalupian through the Tatarian, 267 of 526 (49 percent) of all durably skeletonized marine families became extinct and about 70 percent of the corresponding genera (figure 3.1) (Sepkoski 1992a,b). If genera that are known only from a single stratigraphic stage are removed from the data set (a reasonable precaution since many of them are known only from single localities), the percentage extinction changes to about 72 percent. Recall that no other mass extinction

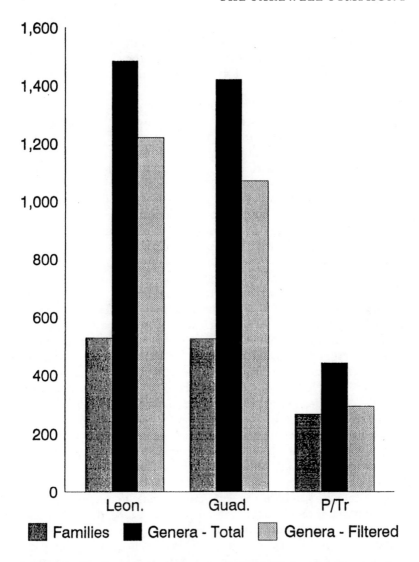

Figure 3.1. Family and generic diversity during the Leonardian, Guadalupian, and at the Permo-Triassic boundary. Generic diversity is shown as both total diversity and filtered by the removal of genera known only from single occurances. Leon = Leonardian; Guad. = Guadalupian; P/Tr = Permo-Triassic boundary. From Sepkoski (1992a).

episode approaches the end-Permian event in magnitude. For comparison, during the end-Cretaceous event about 50 percent of the marine genera became extinct, 57 percent at the close of the Ordovician, and about 48 percent at the end of the Triassic (Sepkoski 1986b).

Ideally we would prefer to know the number of species that disappeared during the extinction, but taxonomic problems, as well as the vagaries of preservation and collecting make this impractical. However, a statistical technique known as rarefaction analysis allows species extinction rates to be estimated from modern generic and family data. In 1979, 894 living echinoid species had been described from 222 genera, 40 families, and 9 orders. If we plot these numbers as a histogram (i.e., the number of genera with a single species, the number of genera with two species, and so on), we produce a frequency diagram with what is called a hollow curve—most genera have only a single species, fewer have two species/genus, etc. From these curves one can calculate the number of species that must be eliminated to remove a given percentage of genera or families. For example, extinction of 50 percent of echinoid species translates to approximately a 22 percent generic extinction and about a 12 percent extinction at the family level (Raup 1979). Obviously we can go the other direction as well, calculating species extinction rates from the generic and familial levels, if we knew the shape of the curve for Permian taxa. However, the shape of these curves has changed as the number of species/family has risen from about six in the Middle Cambrian to ten in the Cretaceous (Valentine 1973a) (although some have questioned whether this really occurred; see chapter 1). To account for this Raup used the living echinoid rarefaction curves, with a species/genus ratio of 4.03; this should give a conservative estimate of the drop in diversity. Using Sepkoski's 1979 data Raup calculated a 96 percent extinction for the end-Permian mass extinction based on the family and order rarefaction curves, and an 88 percent species drop based on the generic curve. Using the same curves but the latest data from Sepkoski I estimated the drop to be about 90 percent using the generic curve, and 95 percent using the family rarefaction curve. The similarity to Raup's 1979 estimate reflects one of the problems of rarefaction analysis, for the estimates of the percentage of generic extinction have changed from 65 percent (Raup 1979) to 78–84 percent (Sepkoski 1989).

These numbers should be treated with extreme caution, however. They are estimates for the entire fauna and assume that all clades had similar rarefaction curves, species/genus ratios, and suffered similar extinction levels. In fact, none of these assumptions is justified. We know, for example, that extinction rates differed greatly between different groups. A single crinoid species (at best, two) survived the extinction, so the percentage extinction was very high. Other groups were only mildly discomforted and suffered much lower species extinctions. Nonetheless, this exercise does provide some general idea of the magnitude of the extinction.

The Evolutionary Faunas described by Sepkoski are, in part, defined by differential survival during the end-Permian mass extinction. The remnants of the Cambrian Fauna declined by about 40 percent while the Paleozoic Fauna (which comprised the bulk of the families) dropped 79 percent, but the Modern Fauna dropped only 27 percent (Sepkoski 1984). Sepkoski notes that the Modern Fauna comprised about 42 percent of total diversity but climbed to 70 percent in the Middle Triassic. The Paleozoic Fauna also suffered higher extinction during the end-Triassic extinction, and this extinction finally ended the dominance of the Paleozoic Fauna. This pattern of selective extinction is closely correlated with mode of life, for most members of the Paleozoic Fauna were sessile, epifaunal suspension feeders. Species with the same habitat in other groups, for example the gastropods, also suffered higher extinction rates than members of the group with different ecologic strategies. This pattern of ecological selectivity provides one of the many clues provided by examining patterns of extinction and survival in individual groups. Other general reviews of end-Permian extinction patterns include Newell (1967b), Hussner (1983), Rhodes (1967), Maxwell (1989), papers in Logan and Hills (1973), and Teichert (1990).

EXTINCTION PATTERNS WITHIN SPECIFIC GROUPS

Foraminifera

THE RECORD of phytoplankton and the zooplankton that feed upon them provides an important record of changes in marine primary productivity. Foraminifera have also received considerable attention because of the extinction of the suborder Fusulinina at the end of the Permian, the only extinct suborder of Foraminifera. Many earlier discussions of foram extinction rates across the boundary (i.e., Tappan and Loeblich 1973; Pitrat 1973; Tappan 1968, 1982) described a lengthy period of decline among fusulinid foraminifera beginning in mid-Permian times. However, this research suffered from inadequate sampling of the latest Permian record, particularly in South China. As more information on this record has become available, it has modified our knowledge of the extinction, concentrating more of the disappearances in the latest Permian.

Family diversity dropped from 40 during the Guadalupian and Tatarian to near 20 in the earliest Triassic; generic diversity declined from just over 100 to about 25 over the same interval, largely as a result of the decline in fusulinids. Fusulinid genera suffered a drop of about 30 percent between the Guadalupian and Tatarian followed by a slight diversification and then continued to decline to the boundary (Tappan and Loeblich 1988; Brasier 1988). According to Sepkoski (1992) all 19 families of the Fusulinina disappeared during the Changxingian Stage; he records the extinction of only a single nonfusulinid family during the stage (figure 3.2). In South China fusulinids were

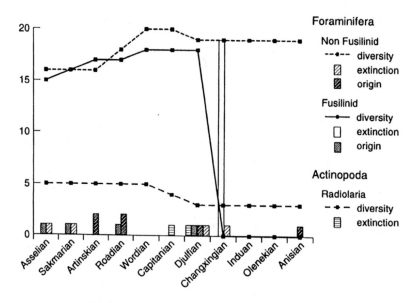

Figure 3.2. Stage-level Permo-Triassic family diversity patterns of Foraminifera and Radiolaria, showing diversity, extinctions, and originations. Data for this chart and subsequent charts in this chapter are from Sepkoski (1992). In all charts only families resolvable to series level were included. Families resolved to series but not stage were apportioned among the stages in accordance with diversity of the other families.

reduced to 9 genera at the base of the Changxingian, 75 percent fewer than during the late Early Permian (Xu et al. 1989).

The nonfusulinid members of other orders do suffer some extinction, but not to the extent of the fusulinids. About 30 percent of the genera belonging to the suborder Textulariina became extinct and 50 percent of the Miolina genera disappeared (Tappan and Loeblich 1988). In South China nonfusulinid forams were abundant during the Changxingian Stage, but only 37 percent of all foram genera survived into the earliest Triassic (Xu et al. 1989).

The differing patterns of extinction reflect two factors. First, fusulinids were largely confined to shallow depths (less than 20 m), while the nonfusuline foraminifera lived in deeper waters, suggesting that extinction was higher in shallow environments than in deeper environments. Second, Brasier (1988) noted that architecturally more complex forms (across all suborders) on tropical shelves were heavily affected.

Sponges

Little data exist on extinction patterns of Permian sponges, but many of the calcisponges are associated with reefs, which were severely affected during the extinction. By Sepkoski's calculations the Demospongia suffered a decline of three families during the Wordian, one in the Capitanian, and two in the Djulfian (figure 3.3). However, overall diversity was only slightly reduced from Lower Permian levels. Among the Calcispongia four families disappeared in the Dorashamian, presumably associated with the disappearance of reef habitats. Hexactinellids showed no decline in family diversity during the Late Permian.

Corals

Tabulate and rugose corals were eliminated by the end-Permian mass extinction, although the (poor) record of nonanthozoan cnidarians does not indicate any changes in diversity. The number of tabulate coral taxa was already far below the Devonian peak, with fewer than 20 genera recorded from the Pennsylvanian through the Permian, and a marked drop in diversity during the Upper Permian (Scrutton 1988). Sepkoski's data record the extinction of six families during the Capitanian and the final five during the Djulfian (figure 3.4). According to Teichert (1990), the youngest tabulates are four genera from the Ali Bashi Formation in the Transcaucus region.

91

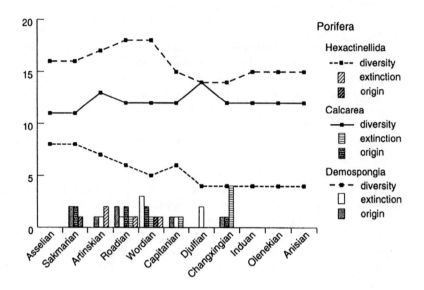

Figure 3.3. Stage-level Permo-Triassic family diversity patterns for the sponges (Phylum Porifera).

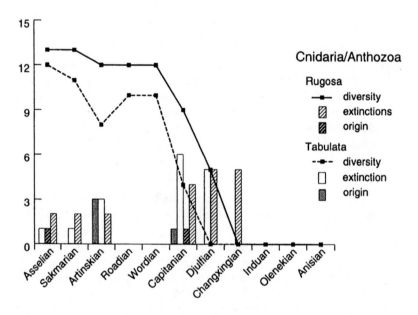

Figure 3.4. Stage-level Permo-Triassic family diversity patterns for the corals (Phylum Cnidaria, Class Anthozoa).

Rugose corals declined throughout the upper Paleozoic. Global diversity was about 60 genera at the close of the Lower Permian (Scrutton 1988) but the group finally disappeared at the close of the Permian with the disappearance of four families in the Capitanian, five in the Djulfian, and five in the Dorashamian (figure 3.4). In South China 10 genera and 39 species have been recorded from the Longtanian (Djulfian) Stage and 6 genera with 30 species in the Changxingian (although 19 of the 30 species were assigned to the genus *Waagenophyllum* (Xu et al. 1989).

Fedorowski's (1989) analysis of biogeographic distribution suggested tabulate and rugose corals were divided into two biogeographic realms, a Cordilleran-Arctic-Uralian Realm and a Palaeo-Tethys Realm. Although the record from the Cordilleran-Arctic-Uralian Realm is less complete than that from the other realm, Fedorowski claims that the corals disappeared in the Cordilleran-Arctic-Uralian Realm by the lower Djulfian. In contrast, and as noted above, a diverse rugose coral assemblage persists into the Changxingian in South China. Fedorowski claims this apparaent diachroneity in extinction is real, and if he is correct it has important implications for the extinction. This pattern suggests that the Tethys region and the Asian microcontinents may have been the final refugia for many groups, and the final extinction in this region proceeded more rapidly than elsewhere. However, considerable attention must be given to problems of correlation, sampling, and other biases before this pattern can be accepted.

The origin of the modern scleractinian corals remains enigmatic, but Oliver (1980) argued convincingly that they were not derived from rugose corals (despite continuing claims to the contrary; see Xu et al. 1989). Differences in septal insertion patterns and skeletal mineralogy strongly suggest that scleractinian corals evolved via the acquisition of the ability to secrete aragonite by a group of sea anemones.

Bryozoans

The geographically widespread bryozoan genera of the Lower Permian gave way to a more provincial biogeographic pattern during the mid-Permian, largely associated with the development of a diverse, distinctive Tethyan bryozoan assemblage (Ross 1978). The bryozoan extinction record suggests a gradual decline beginning well before the end of the Permian. Global family diversity dropped 71 percent, from 24 families in the Wordian: 10 families disappear in the Capitanian,

5 in the Djulfian, and 3 in the Dorashamian. Another 4 families have last appearances during the Induan (figure 3.5). At the generic level, about 52 genera (68 percent) are last recorded from the Guadalupian; 14 more (64 percent) from the Tatarian (Ross 1978). An unpublished generic database compiled by Robert Anstey supports Ross's conclusions with extinctions of 14 genera during the Roadian, 23 genera during the Wordian, 23 during the Capitanian, but only 13 during the Tatarian. As always, the problem of biased preservation is real, but Late Permian bryozoan faunas are largely restricted to Tethyan regions (Ross 1978; Sakagami 1985). Few are known from deposits of latest Permian age, either in western Tethys or South China. For example, in the Transcaucus region bryozoan diversity drops markedly in the Guadalupian Stage. Bryozoan colonies are abundant in the Djulfian but only 5 genera and 6 species have been found; all are extinct by the end of the lower Djulfian (Ruzhentsev and Sarycheva 1965). Similarly, in Kashmir at Guryul Ravine bryozoan diversity drops from 20 species in 15 genera in Member A to 8 species and 7 genera in members C and D just below the boundary (Nakazawa and Kapoor 1981). The remarkable paucity of latest Permian bryozoans

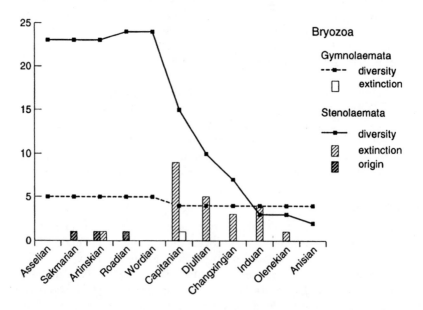

Figure 3.5. Stage-level Permo-Triassic family diversity patterns for the Phylum Bryozoa, classes Gymnolaemata and Stenolaemata.

strongly suggests that most disappearances in this clade occurred well before the end of the Permian, many during the Guadalupian Stage.

A final consideration is the relationship between the Paleozoic bryozoan orders and the post-Paleozoic orders. The traditional view has been that four stenolaemate orders became extinct (the Trepostomata, Cryptostomata, Cystoporata, and Fenestrata), while the stenolaemate Order Cyclostomata and the gymnolaemate Order Ctenostomata continued on into the Triassic, with the latter eventually giving rise to the Cheilostomata. More recent work, based on new studies of Triassic bryozoans and new systematic treatments, suggests that only the Order Fenestrata became extinct at the end of the Permian (Taylor and Larwood 1988). Although the relationships remain to be worked out, under either of the possible scenarios advanced by Taylor and Larwood, the other three stenolaemate orders became extinct during the Mesozoic and the majority of post-Paleozoic bryozoans are derived from two orders that had been minor elements of late Paleozoic bryozoan assemblages. Of these latter two orders Taylor and Larwood (1988:111) emphasize that "their evolutionary pattern fits the classic picture of initially subordinate, morphologically simple groups surviving mass extinction events (Permian and Triassic) that brought about the demise of previously dominant, complex groups, and subsequently radiating into vacant habitats."

Brachiopods

Brachiopods are perhaps the most important group of upper Paleozoic fossils in terms of both diversity and abundance. A number of odd-looking articulate brachiopod groups evolved during the Permian, including the cone-shaped, reef-dwelling richthofenids, spiny productids, and the strange, lobate oldhaminids. Inarticulate brachiopods were unaffected by the extinction, but the articulate brachiopods suffered an irreversible decline in diversity and abundance. According to the traditional taxonomic scheme, the orthids became extinct at the close of the Permian, the spiriferids straggled across with but lost no superfamilies, the strophomenids were virtually wiped out, while the rhynchonellids and terebratulids were virtually unaffected. Global family diversity declined by 55 families (90 percent) from the Roadian to the earliest Triassic, with 77 percent of the drop occurring during the Djulfian and Dorashamian (figure 3.6). According to Carlson (1991), 10 of 23 superfamilies, and perhaps as many as 95 percent of all genera, became extinct.

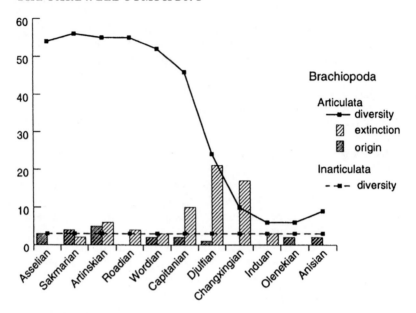

Figure 3.6. Stage-level Permo-Triassic family diversity patterns for the Phylum Brachiopoda, classes Articulata and Inarticulata.

The record across different boundary sections exhibits considerable variation in the diversity of the assemblages. The fauna in South China is far more diverse than in western Tethys (other than Greece) and appears to remain so until just below the boundary (although the most diverse Late Permian brachiopod assemblage is found in Nepal; see chapter 2). However, the actual diversity of Changxingian brachiopods is unclear, since the fauna has not been completely described. According to Yin and colleagues (Yin, Xu, and Ding 1984, quoted in Xu et al. 1989), 98 species and 46 genera are present with only 17 species in 9 genera (17 percent) persisting into the earliest Triassic. Liao (1980) cites 60 genera and 130 species. The mixed faunal assemblages of the earliest Triassic include 10 genera with 28 species but are followed by large numbers of the inarticulate brachiopod *Lingula*.

Articulate brachiopods are less diverse in the Transcaucus sections. Ruzhenstsev and Sarycheva (1965) report 91 species across 48 genera. Thirty of these species are restricted to the Guadalupian. In comparison they note 281 species were recorded from Upper Permian reefs in the northern Caucusus, although the age (Guadalupian, Djulfian, or Dorashamian) is not recorded. Over half of the Transcaucus species are endemic. The Djulfian assemblage (actually Djulfian plus Dorashamian in current usage) includes 54 species and the lower Djulfian is considerably more diverse than the underlying Guadalupian assemblage. Most species disappear well before the boundary and only 9 species are recorded from the Induan Stage. As in South China, two different facies assemblages are present, a clay-bottom assemblage and a higher-energy, carbonate-rich assemblage. The brachiopods that persist into the Triassic come largely from the clay-bottom assemblage and are fairly small. At Guryul Ravine in Kashmir brachiopod diversity drops from 18 species in 14 genera in the lower Zeewan Formation to 2 species in 2 genera at the boundary, and then to 0 (Nakazawa and Kapoor 1981).

The traditional view suggests that the pattern of extinction is highly uneven. Twenty families and four to five superfamilies assigned to the Strophomenida disappeared, yet this order had the highest familial diversity during the Permian of any group of articulate brachiopods. In contrast, all eight superfamilies of spiriferids survived the extinction, but seven of the eight died by the end of the Triassic. Of the 13 articulate brachiopod superfamilies only a single family actually crosses the boundary in 11 cases, and in most cases familial continuity involves the extinction of an Upper Permian genus and the origination of a new, Triassic genus. This is true of the rhychonelloids, terebratuloids, and other groups.

However, many of these higher-level groups are not monophyletic clades but are paraphyletic assemblages that do not include all of the descendants of a clade, some of which may have been hived off into other, paraphyletic assemblages because of apparent differences in morphology or other attributes, as discussed in chapter 1. If considerable paraphyly exists, apparent taxonomic selectivity may be an artifact while real selectivity is obscured.

Fortunately, Sandy Carlson has analyzed the extent of this problem as part of her ongoing work reconstructing the phylogenetic relationships among the major brachiopod groups. In her analysis, Carlson (1991) found that pentamerids, atrypids, and terebratulids are monophyletic while the rhynchonellids, strophomenids, and orthids are paraphyletic; the spiriferids are divided into a number of discrete clades. Unlike the taxonomic patterns described above, there is no coherent pattern to the extinction of monophyletic clades (figure 3.7). The difference between the taxonomic and phylogenetic analyses of extinction patterns suggests not only the level of disparity between the two approaches but the difficulty in simply using taxonomic

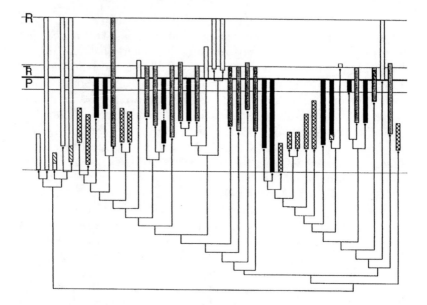

Figure 3.7. Cladogram of major articulate brachiopod groups. Shadings show groups traditionally united into a single superfamily. If these groups were monophyletic they would be united on the cladogram. The fact that they are largely broken up into different clades suggests that few of the traditionally recognized superfamilies are monophyletic clades. From Carlson 1991; copyright by Dioscorides Press.

structure as an indicator of phylogeny. Since the spiriferids are not a clade, their apparently greater extinction, particularly relative to the strophomenids, is not real. Hence the classic view that they suffered greater extinction across the Permo-Triassic boundary is incorrect. Whether the presence of spiralia (the feature uniting these various clades into the spiriferid grade) is associated with increased extinction remains unclear. However, the importance of rigorous phylogenetic analyses in discerning patterns of extinctions susceptibility should be obvious.

The phylogenetic analysis of the articulate brachiopods allows a character-based approach to studying the functional, ecologic, and biogeographic patterns of extinction and survival (Carlson 1991). The only significant pattern among the characters Carlson analyzed was an increased extinction rate among brachiopods with noninterlocking hinge structures and a faster recovery after the extinction of groups with interlocking hinge structures. Carlson plausibly suggests that brachiopods with an interlocking hinge structure have greater eco-logic variability—certainly all living brachiopods have this structure.

Carlson's analysis is a preliminary study of extinction patterns within the brachiopods, but the rigorous use of phylogenetic analysis in identifying patterns of phylogenetic and functional selectivity points the way for future work across all clades. The cautionary tale of the spiriferids and the strophomenids and the lack of similar stud-ies in other clades may mean that we know less about the patterns of extinction and survival than we choose to believe.

Molluscs
Gastropods

Gastropods experienced less extinction during the end-Permian than any other clade except the bivalves. Because of the large number of ecological roles they fill, their broad environmental distribution, and high survivorship, Permian gastropods are one of the best groups in which to study the factors influencing extinction and survival. Clades with low extinction rates are more likely to yield useful information on these factors than groups with a high percentage of extinction. Overall family extinction was about 20 percent (figure 3.8); no major superfamilies of gastropods became extinct, although the Bellero-phontina, Euomphalina, and Subulitina all began declines that cul-minated in their extinction in the Triassic. Two minor superfamilies appear to have become extinct: the platyceratids disappeared with the dramatic reduction of their hosts, the crinoids, and the Pseudopho-roidea, an odd group of presumed mesogastropods also disappeared. I say appear to have become extinct because Ponder and Waren (1988) claim that the platyceratids and the pseudophorids are Paleozoic ancestors of the modern Neritopsina and Xenophoroidea, respec-tively. Resolution of this puzzle will require additional phylogenetic studies. The Pleurotomariina, the most diverse gastropod superfamily during the late Paleozoic, declined drastically during the extinction, but rebounded quickly thereafter.

The first detailed discussion of lineages that disappear in the Late Permian only to reappear in the Middle Triassic came in a paper in 1973 written by Roger Batten, where he noted that latest (Djulfian and Dorashamian here) Permian gastropod faunas were exceedingly depauperate, with only 26 genera known. The Chhidru Formation had the most diverse gastropod assemblage, with 14 genera. Most Upper Permian genera are long-ranging with broad environmental tol-

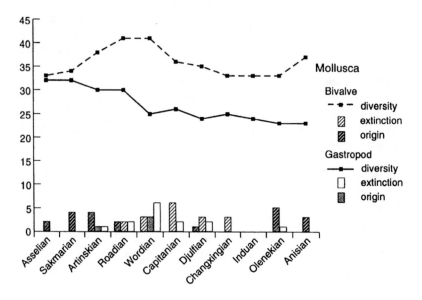

Figure 3.8. Stage-level Permo-Triassic family diversity patterns for the Phylum Mollusca, Classes Bivalvia and Gastropoda.

erances (Batten 1973; Erwin 1989b). Batten noted that many gastropods seem to be associated with hypersaline conditions, but it may be more accurate to emphasize their broad environmental range rather than adaptation to a specific environmental condition. Most importantly, Batten (1973:603) noted: "It is quite obvious that the Dzulfian and Scythian faunas, as now known, lack genera which occur in the Guadalupian, *and* reappear in the Ladinian! There are 32 genera and 16 families of Guadalupian age in the Triassic, *none* of which are known in the Dzulfian." Jablonski (1986a) subsequently described these genera as "Lazarus taxa": species and genera that disappear from the record only to reappear when conditions return to normal. The gastropods have a large number of Lazarus taxa, although the phenomenon is also widespread among bivalves and articulate brachiopods.

In a recent analysis of the role of trophic level during both background and mass extinction intervals, Allmon et al. (forthcoming) found that suspension-feeding gastropods generally had higher extinction rates during both background times as well as during the end-

Ordovician and end-Devonian mass extinctions. However, this pattern breaks down during the end-Permian extinction when all trophic groups show increased susceptibility. Detritus feeders remain the most extinction-resistant, as usual.

Bivalves

Bivalves were relatively unaffected by the extinction. Six families disappeared during the Capitanian, three during the Djulfian, and three during the Dorashamian, but with one new family during the Djulfian this amounts to only a 26 percent drop (figure 3.8). The epifaunal Pterioida were heavily affected, as were the Pholadomyoida, although the latter were largely restricted to Australia, Siberia, and South America. The reduction in the Pterioida was permanent, but no other clade or life habit experienced an irreversible change in fortune as a result of the extinction. At best, bivalves experienced only a minor interruption in long-term evolutionary trends (Hallam and Miller 1988).

The Late Permian bivalve assemblages in many areas are fairly depauperate; only in South China is the fauna diverse (Teichert 1990). Yin (1985a) noted the presence of 50 genera and 112 species in the Changxingian, of which 21 genera and 56 species are pectinids. In South China 43 of 54 (85 percent) bivalve genera disappeared during the late Changxingian, 18 genera are known from the overlying Triassic, 10 of which are new. As was the case with the gastropods, the actual magnitude of the extinction is magnified by poor taxonomy. For example, the Permian genus *Pernopecten* is a synonym of *Entolium* (Yin 1985a). Yin argues that there is no evidence of a sudden extinction among the bivalves, but rather a gradual replacement of older forms by newer ones.

In contrast to the highly provincial nature of Late Permian bivalve assemblages, Early Triassic bivalve assemblages tend to be highly cosmopolitan. Furthermore, many "Triassic" bivalve genera actually first appeared in the Late Permian. While many new genera do appear in the Early Triassic, pteriomorphs (scallops) and groups with eulamellibranch gills are absent and do not reappear until the Middle Triassic (Yin 1985a; Nakazawa and Runnegar 1973). These Lazarus taxa may amount to two or three times the recorded diversity of Early Triassic bivalves (Nakazawa and Runnegar 1973), further reducing the severity of the extinction for the bivalves.

Cephalopods

Cephalopods were among the most severely affected groups, with 22 families disappearing between the Wordian and the boundary, for an overall decline of 50 percent in diversity (figure 3.9). Different groups were affected in very different ways, however. Orthocerid diversity had already dropped to 3–4 described genera by the Late Permian (Teichert 1990). Nautiloids, on the other hand, were reasonably diverse but apparently unaffected by the extinction. Such was not the case among the ammonoid cephalopods, although the pattern was more one of gradual change than catastrophic transition (Wiedmann 1973). Goniatites were wiped out while the ceratites, which originated in the Roadian, suffered only minor extinction near the boundary. The Otoceratidae and Xenodiscidae (which together gave rise to all Triassic ammonoid families) had already appeared by the Late Permian and diversified explosively during the Early Triassic (Tozer 1980). The small order Prolecanitida is not known from the Changx-

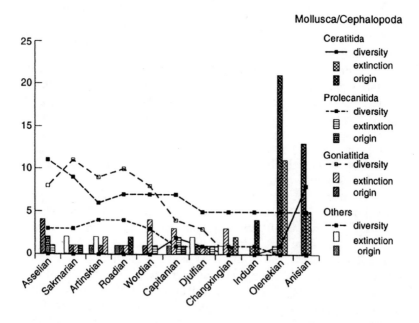

Figure 3.9. Stage-level Permo-Triassic family diversity patterns for the major groups of cephalopods (Phylum Mollusca, Class Cephlopoda).

ing Formation but reappears briefly in the Lower Scythian before becoming extinct (Teichert 1990; Tozer 1980).

In the Changxingian of South China 154 species and 33 genera of ammonoids have been recorded and most occur in a siliceous facies. A total of 96 ammonoid species disappeared at the end of the Changxingian, giving a 62 percent extinction, although Xu et al. (1989) cite a 98 percent extinction. The bulk of the Changxingian genera are ceratids of the superfamily Xenodiscaceae, imparting a decidedly Triassic aspect to the assemblage (Teichert 1990).

Among the other molluscan classes, there are no recorded extinctions at the family level among the polyplacophora (chitons) or scaphopods. The "monoplacophora" disappear from the fossil record, only to be rediscovered living in the deep sea in 1953, and the hyolithids (which may not be molluscs at all) finally became extinct during the Guadalupian. All of these groups had fairly low diversity through the late Paleozoic.

Echinoderms

The Paleozoic-Mesozoic boundary has long been viewed as a major crisis in the history of echinoderms, with the disappearance of the major Paleozoic groups, particularly the stalked crinoids and the blastoids (figure 3.10). As Bottjer and Ausich (1986) noted, one of the effects of the extinction was the disappearance of the epifaunal tier: most attached epifaunal organisms became extinct and there were likely substantial changes in the maximum height of epifaunal tiers during this interval. Unfortunately, as they note, the record of epifaunal echinoderms during this interval is relatively poor, making it difficult to gauge the pattern of change in epifaunal tiering. During the subsequent recovery the composition of echinoderm assemblages changed, with echinoids taking a far more prominent role than they had during the Paleozoic. The claim that echinoderms were almost completely wiped out by the end-Permian mass extinction considerably exaggerates the importance of the extinction relative to long-term trends within the phylum. Echinoids, blastoids, and crinoids were all fairly rare throughout the Permian and while each group did suffer at the close of the period, none was a prominent component of faunas immediately preceding the extinction.

Echinoid diversity was not only low throughout the Permian but the same clade, the cidarids, accounts for the bulk of both Late Permian and Early to Mid-Triassic echinoids (Kier 1973). Only two gen-

103

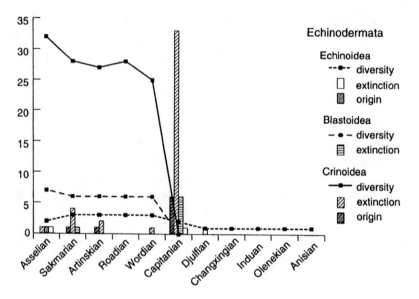

Figure 3.10. Stage-level Permo-Triassic family diversity patterns for the major Permian echinoderm classes, the echinoids, blastoids, and crinoids.

era survived the extinction (Kier 1984). Echinoid diversity underwent an extreme bottleneck during the extinction.

The crinoids nearly disappeared as well, with the extinction of 30 families during the Capitanian, including the subclasses Camerata and Flexibilia; three suborders also disappeared. Most of these families are known only from the rich faunas of Timor, which are likely Guadalupian, although there is some uncertainty about the age. The only record of Tatarian crinoids are crinoid stems (ossicles) from the Ali Bashi Formation of Iran (Teichert and Kummel 1973) and South China (Reinhardt 1988b). Some crinoids obviously survived the extinction, although the relationship between the Paleozoic and post-Paleozoic crinoids is poorly understood. Post-Paleozoic crinoids are assigned to the subclass Articulata, so the crinoids, like the echinoids and asteroids, are a classic example of a group that suffered a severe bottleneck during the extinction and was strongly modified as a result. Schubert, Bottjer, and Simms (1992) described articulate crinoid ossicles assigned to *Holocrinus smithi* from the late Scythian Moenkopi Formation of Utah.

Blastoids had begun their terminal decline well before the end of the Permian. According to Kier (1973), only 4 genera are known from the Pennsylvanian, and of the 23 Permian genera 16 are from the Timor fauna in the Indonesian archipelago. These are probably the youngest blastoids, suggesting the group disappeared during the Guadalupian. Teichert (1990) notes the possible presence of a single genus in the early Djulfian.

Existing asteroid (starfish) orders have generally been traced back to the Ordovician, but no families are known to cross the Permo-Triassic boundary. Gale (1987) demonstrated that post-Paleozoic asteroids form a single monophyletic group differing substantially from known Paleozoic forms. Flexible arms, suckered tube feet, and the ability to evert the stomach (key to the feeding strategy of modern asteroids) appeared in post-Paleozoic asteroids, greatly increasing the ecologic range of the group. The phylogenetic pattern suggests that the extinction may have been involved in the transition, but the record of the class is too poor to know for sure. The record of ophioroids (brittlestars) is too poorly understood at present for discussion (Paul 1988).

Arthropods

The importance of the end-Permian extinction to marine arthropods may be gauged by the fact that Briggs, Fortey, and Clarkson (1988) do not even mention the event (figure 3.11). Trilobites finally disappeared at the end of the Permian, but were already limited to three small families of Ptychoparids by the beginning of the Permian. Perhaps surprisingly, according to Fortey and Owens (1990) three of their eight morphologic classes of trilobites are represented by Upper Permian species. Trilobites are not known from the Iranian or Armenian sections (Teichert 1990), but *Pseudophillipsia* and *Acropyge* have been recorded from the Changxingian Formation of South China (Zhao et al 1981).

There are scattered disappearances at the family level among other arthropod groups but the only diverse marine arthropod group during the Permian were the ostracodes. Ostracode family diversity began to decline in the Wordian and continued to drop through the Induan. Ten families (58 percent) disappeared before the boundary, three more during the Induan Stage.

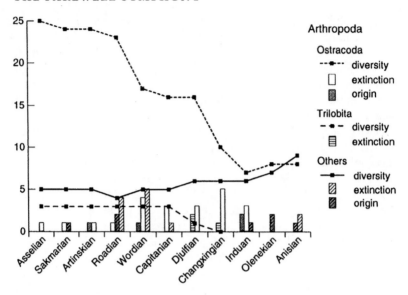

Figure 3.11. Stage-level Permo-Triassic family diversity patterns for the major Permian marine arthropods, the osatracods, trilobites, and all others.

Conodonts

Conodonts passed through the end-Permian mass extinction with seeming indifference, as Sweet (1973) noted. In South China there is no decline in species diversity, although the number of specimens found is low. The group is of particular importance for global correlations in the Permian and Triassic, in part because of the relative extinction-resistance of the clade. Worldwide, 8 genera and about 52 species are known from the Late Permian and 9 genera and 40 species from the Scythian (Clark 1987), although Clark notes that there were probably never more than 8 species alive at any one time. The group experienced rapid diversification in the earliest Triassic.

Vertebrates

In his review of fish diversity, Thompson (1977) observed that total fish diversity reached a peak in the Late Devonian and Early Carboniferous and declined steadily until morphologic innovations during the Triassic rejuvenated the group. An extensive diversification occurred during the Triassic, but this may in part reflect an increased opportunity for preservation in Triassic sediments relative to the Permian (Schaeffer 1973). Thompson concurred with Schaeffer's assessment that the mass extinction had a negligible impact on fish diversity. Yet unlike Schaeffer and Thompson, who considered both marine and nonmarine groups together, Pitrat (1973) considered the nonmarine families separately from exclusively marine families. He uncovered a pattern of selectivity, with far higher extinction rates among the marine fish than among the nonmarine groups. However, the Late Permian fossil record of several nonmarine groups, including the lungfishes (Dipnoi), Osteolepiformes, Porolepiformes, and Actinistia (the latter three formerly known as the crossopterygii; see Benton 1990b) is so scanty as to provide little information.

SYNTHESIS

THE VARIETY of apparent extinction and survival patterns suggests something of the complexity of the end-Permian mass extinction. Some groups (conodonts, nonfusulinid foraminifera, bivalves, nautiloid cephalopods, and bellerophontid gastropods) endured the event in blissful ignorance. Several groups appear to have been in decline well before the boundary and became extinct by the end of the Capitanian, including blastoids and camerate crinoids and tabulate corals. Other clades were in decline but managed to survive into the Tatarian at greatly reduced diversity, including the trilobites, bryozoa, rugose corals, and perhaps crinoids. Other clades, including the fusulinid foraminifera, articulate brachiopods, and ammonoid cephalopods were quite diverse during the Dorashamian but experienced severe constrictions at the close of the period. Finally, other groups were less diverse during the Late Permian and suffered variable levels of extinction only to recover quickly in the Triassic, including the nonbellerophontid gastropods, the cidaroid echinoids, sponges, and the rhynchonellid, spiriferid, and terebratulid brachiopods.

What general conclusions can we draw from these descriptions of pattern about the rate and mechanisms of extinction? Three points seem clear. First, the diversity of many groups appears to have declined much earlier outside of South China than within South China itself, at least in the corals, fusulinids, brachiopods, bivalves, and cephalopods. This could indicate that South China preserves the best record of the extinction and that the magnitude of the regression obscures the record elsewhere, or that South China represents a temporary refugium and many species persisted longer in South China than elsewhere. The latter possibility is strengthened by the isolated location of South China in the Late Permian and Sweet's (1992) conodont correlation of the Changxingian with the Griesbachian (earliest Triassic) of western Tethys (chapter 2). Sweet's correlation suggests that the extinction may have occurred later in South China than elsewhere, but it has not been completely accepted by other biostratigraphers working on the Permian. If the carbon-isotopic shift is a global signal, it suggests that the extinction was globally correlative.

Second, there is considerable variation in the quality of the data analyzed. Carlson's (1991) phylogenetic work on articulate brachiopods, and Taylor and Larwood's (1988) on bryozoans all suggest that the magnitude of the extinction has been considerably exaggerated by poor taxonomy, a point Batten made in 1973 with reference to gastropods. Cleaning up these problems should reduce the apparent severity of the extinction somewhat but is unlikely to change our view of the end-Permian mass extinction as the most severe of the Phanerozoic. More importantly, it will allow paleontologists to assess the patterns of extinction and survival more accurately, recognizing, for example, that there was no differential extinction of strophomenid brachiopods relative to spiriferid brachiopods and allowing a more fruitful inquiry into patterns of functional, ecological, environmental, and biogeographic control of selectivity. This problem of inadequate taxonomy may be most severe in South China, which contains some of the most important Late Permian sections. Unfortunately, despite major efforts on the part of many Chinese paleontologists, much of the fauna remains either undescribed or described only in Chinese-language publications, and largely unavailable to the rest of the paleontologic community. Additionally, the quality of the systematic work is highly uneven; while much of the systematic work is of the highest quality, the quality of publication (paper, illustrations, etc.) is, unfortunately, generally inferior. In other cases the systematic treatments are so unreliable as to be an uncertain guide to extinction patterns.

Finally, even after the systematic and phylogenetic problems are corrected, how much confidence should we place in these apparent patterns of extinction? The use of disappearance and apparent extinction at the beginning of this section was intentional. The quality of the fossil record is notoriously poor in many instances (although it may well be far better than often assumed). Does the fossil record accurately reflect the actual timing of extinction? The abundant Lazarus taxa among the gastropod, articulate brachiopod, and bivalve genera suggest something of the problem. Does this mean that many bryozoans, for example, simply disappeared from the record during the Guadalupian but did not become extinct until the Changxingian? If so, the apparently prolonged extinction was actually far more rapid. Problems of preservation and sampling bias tend to produce a gradual or stepwise pattern of extinction even for a catastrophic, virtually instantaneous mass extinction. These problems are addressed in the following section.

SAMPLING PROBLEMS

THERE ARE a variety of biases that reduce the accuracy of the fossil record. One is the effect of variation in sample size. For example, until recent collecting, the decline in ammonite diversity in the Late Cretaceous closely paralleled the reduction in the areal extent of rock deposited toward the close of the period (Kennedy 1977; Signor and Lipps 1982). Does this decline in ammonite diversity reflect true extinction or only sampling problems? In plotting patterns of diversity paleontologists assume the patterns are real. Over the past two decades increasing attention has been paid to the validity of this assumption and the problems of sampling bias (see particularly Raup 1976).

A related problem is that of artificial range truncations during extinction events, now known as the Signor-Lipps Effect, following a seminal paper on the phenomenon by Phil Signor and Jere Lipps. They were particularly concerned with whether an apparently gradual pattern of disappearance actually required a gradual extinction or whether it was equally consistent with a catastrophic extinction. Since the likelihood of actually preserving and then recovering the last specimen of any species, and thus the true extinction point, is virtually nil, the recorded range of a species will always be less than the true range (ignoring the effects of reworking). The length of the gap between the apparent range and the true range varies with the type of organism, the environment in which it lived, and other factors.

Signor and Lipps realized that if large numbers of species are considered, the extent of artificial range truncation should vary randomly between species. They modeled the process under various assumptions and demonstrated that as sampling improves the apparent diversity curve progressively approaches the true diversity curve. Signor and Lipps also noted that marine regressions will enhance the effect of artificial range truncation, pushing the apparent extinctions back to a time where there are sufficient deposits to sample. They concluded: "If regressions are gradual, they could have the effect of amplifying the apparent gradual nature of the extinctions, simply by reducing sample size as the regression progresses, at least in continental sections" (Signor and Lipps 1982:294). Thus the best record of true extinction patterns may occur in groups with a more complete record, such as planktonic organisms. A study of sampling bias in the Gulf of California assumed that a mass extinction occurred today, then examined the stratigraphic record of species in sediment cores to determine the pattern of recorded disappearances (Meldahl 1990). This study confirmed the Signor-Lipps Effect, but demonstrated that the use of last occurrence and stratigraphic abundance data potentially allows paleontologists to distinguish between sudden, stepwise, and gradual extinction patterns.

Other analyses of the effects of sampling bias on diversity studies (Koch 1987; Koch and Morgan 1988) demonstrate the importance of the problem and provide a series of tests that can be applied to diversity data before one proposes an explanation for a pattern that does not actually exist. Additionally, Marshall (1990) developed a technique to assign confidence intervals to the known ranges of a species using only the length of their stratigraphic range and the number of fossil horizons from which they have been recovered. Application of this technique to species ranges near a mass extinction allows some estimate of the confidence one should place in the apparent record of extinction.

How does this work apply to the end-Permian mass extinction? The simple fact is that the regression and the limited number of latest Permian sections (particularly outside of South China) greatly increases the probability that sampling problems have severely altered the actual pattern of extinction. Unfortunately, no studies of the actual effect of sampling problems have been published to date, so we really do not know how much of a problem we have. Clearly, prudence cautions against accepting at face value the apparent extinction patterns described earlier in this chapter.

PATTERNS OF TAXONOMIC AND ECOLOGIC SELECTIVITY

MANY OF the selective extinction patterns have been identified in the previous section: nearshore taxa, sessile forms, filter feeders, and taxa outside Tethys all appear to have suffered higher extinction rates than other groups, although in many cases the evidence is scanty at best. New systematic studies, biogeographic analyses, and fieldwork are clearly required to provide more substance to these apparent patterns. Other selective extinction patterns have been suggested as well, and will be discussed in this section.

Reefs

Tropical regions in general and reefs in particular are frequently assumed to be among the most severely affected ecosystems during mass extinctions (Jablonski 1986b; Raup and Boyajian 1988), but there is little evidence to support this canard. The distinctive array of Carboniferous and Permian reefs was dominated by sphinctozoan sponges, bryozoans, stromatoporoids, and calcareous algae. Missing from all late Paleozoic reefs are extensive framework builders, typified by scleractinian corals today (Wilson 1975; James 1983; G. D. Stanley 1988). This reef assemblage disappeared at the close of the Permian. Seven different types of organic buildup, including reefs, have been identified in the Permian: calcisponge-algal reefs, *Tubiphytes*/algal crust reefs, stromatolitic reefs, bryozoan/algal reefs, *Palaeoaplysina* reefs (an odd organism, either a sponge or stromatoporid), phylloid algal buildups, and rugose coral reefs, the only one dominated by corals (Flugel and Stanley 1984). Although the last two reef types are restricted to the Early Permian, the others all occur in the Late Permian; only the calcisponge/algal reefs and the *Tubiphytes*/algal crust reefs reappear in the Triassic. Latest Permian *Tubiphytes*/algal reefs and calcisponge/algal reefs in South China and on the Greek island of Skyros are similar to Early Permian and early Late Permian reefs elsewhere (Flugel and Stanley 1984; Flugel and Reinhardt 1989). Reefs are absent from the Early Triassic but Middle Triassic reefs are overwhelmingly composed of holdovers from Permian reef communities, particularly the association of foraminifera, calcisponges, *Tubiphytes*, and algae. Other Permian holdovers are the calcisponge *Girtyocoelia* and phylloid algae (Flugel and Stanley 1984; G. D. Stanley 1988).

Uppermost Permian reefs in Skyros, Greece and South China were very diverse and show no indication of gradual extinction (Flugel and Reinhardt 1989). Flugel and Reinhardt note that algal limestones persist after the disappearance of the Tudiya bioherm in Sichuan, China, for example, indicating that warm shallow-water marine conditions continued, although without reefs.

The transition from late Paleozoic to Triassic reefs occurred in response to the appearance and differential diversification of scleractinian corals and other new groups in the Middle Triassic, following the reestablishment of basically Permian-type reefs. Sheehan (1985) argues that the lack of Early Triassic reefs reflects the complexity of ecological interactions present on reefs and the time necessary to recreate these connections. In contrast, G. D. Stanley (1988) points out that modern reefs return quite quickly following ecological disturbances (although this involves only local immigration). Since the evolution of new groups appears to have been relatively unimportant to the establishment of Triassic reefs, the lag suggests that environmental perturbations associated with the mass extinction persisted through the Early Triassic. He notes that the recovery of reefs was rapid once favorable environments reappeared.

Reefs are largely confined to tropical latitudes, leading to the generalization that the tropics suffer more severely during mass extinctions than high- or low-latitude regions. Jablonski (1986b) examined the record of articulate brachiopods across the Permo-Triassic boundary and confirmed that while 75 percent of the exclusively tropical families became extinct only 56 percent of the extratropical families did. Similar studies have not been performed for other groups but, for gastropods at least, are complicated by other factors. Most Permian gastropod families are either geographically widespread or endemic, largely to tropical regions. Thus exclusively tropical families might have suffered increased extinction because they were tropical or because they had more restricted geographic distribution.

Developmental Mode

There is suggestive (although not yet persuasive) evidence that marine invertebrates with planktotrophic larval development may have experienced greater extinction during the end-Permian mass extinction than did species with nonplanktotrophic development. The larvae of species with planktotrophic development feed on plankton and thus require smaller eggs than do the nonfeeding, nonplanktotrophic species, which must be supplied with yolk as a food source. Consequently, planktotrophic species can produce a large number of eggs and the longer duration of these floating larvae allows broader geographic distribution, more widespread species, and lower rates of speciation and extinction.

Fortunately, since we obviously cannot examine the developmental mode of Permian species directly, indirect evidence is available. Although modern crinoids and archaeogastropods (exclusive of the neritaceans, which really are not good archaeogastropods anyway) are nonplanktotrophic, there is strong evidence that their ancestors were planktotrophic. Planktotrophic larvae are dependent upon a steady supply of plankton and predominate in warm, shallow, low-latitude waters today. Articulate brachiopods were one of the dominant members of late Paleozoic benthic communities, particularly in tropical regions, yet modern articulate brachiopods are nonplanktotrophic and are more common in temperate and high latitudes than in the tropics or subtropics, leading Valentine and Jablonski (1983) to suggest that Permian brachiopods may have been planktotrophic. Crinoids and articulate brachiopods and, to a lesser extent, archaeogastropods suffered particularly heavy extinction, suggesting that conversion to largely nonplanktotrophic development in these groups occurred by selective extinction in the Late Permian (Strathman 1978; Erwin and Valentine 1984; Valentine and Jablonski 1983, 1986; Valentine 1986). In addition, Valentine (1986) noted that orders of durably skeletonized marine invertebrates and families of bivalves and gastropods that originated during the Early and Middle Triassic are overwhelmingly nonplanktotrophic, while those that originated in the Late Triassic are largely planktotrophic. He suggests that this pattern is consistent with a lengthy period of unfavorable conditions for planktotrophic organisms, through at least the Early Triassic.

While the possibility of selective extinction of planktotrophic developers is attractive, the quality of the data supporting this hypothesis is not high and considerably more work must be done to place the hypothesis on a firmer footing. Furthermore, it is far from clear that planktotrophic development, per se, was the basis of the selectivity. Planktotrophy is associated with a variety of other characteristics. Planktotrophs predominate in nearshore environments, which appear to have suffered more severely than offshore regions. Thus the preferential extinction of planktotrophs could reflect increased extinction in the nearshore and/or the tropics rather than a drop in the plankton required for feeding by planktotrophic larvae.

At the same time, it is worth noting that the heightened severity of the extinction of articulate brachiopods and crinoids may reflect a combination of reduction in available plankton and increased near-shore extinction.

The data presently at hand suggest that the marine extinction began, at least in some groups, by the close of the Capitanian and accelerated during the Djulfian and Changxingian. The overall quality of the data is not high however, and sampling biases have almost certainly exaggerated the length of the extinction and systematic problems have increased the apparent magnitude. Nonetheless, it does not appear that either problem is sufficiently pervasive to alter the general pattern. The extinction was highly uneven. Benthic, sessile, filter-feeding groups seem to have been the most heavily affected, with shallow-water taxa, species-poor genera, and geographically restricted taxa all more heavily affected than more broadly distributed groups. Several authors have suggested that many of the survivors were drawn from the ranks of salinity-tolerant forms but there is little support for this conclusion, although survivors appear to have had, in general, broader environmental tolerances than genera that became extinct. Many groups were already dwindling in both diversity and abundance by the mid-Permian, others remained diverse, at least in South China, until near the end of the period. As noted earlier, the high diversity in South China raises an important question: does this accurately reflect global diversity patterns, does South China represent a refugium, albeit temporary, for many taxa, or does it simply reflect the widespread latest Permian outcrops in South China? The answer to this question has important implications for the pace of the extinction and in our search for the mechanisms behind it.

The avenues for future research suggested by this chapter are many and varied. Clearly we need considerably more refined data on the pace of disappearances based on well-controlled sampling schemes through the Late Permian and Early Triassic. In turn, analyzing such data requires resolving the many correlation problems discussed in the preceding chapter. Finally, a considerably amount of taxonomic and phylogenetic work is needed to improve the quality of the systematic data.

IV

Pelycosaurs, Pteridosperms, and Palaeoptera

JUST AS the patterns of extinction and survival in the oceans provide important clues to changes in climate, ocean chemistry, and diversity during the marine phase of the end-Permian mass extinction, the history of life on land preserves similar information. Moreover, if little happened on land our search for the causes of the extinction narrows to changes that affect the sea. On the other hand, a correlative extinction in the terrestrial realm precludes mechanisms that act solely in the oceans. The worst of all possible worlds is an ambivalent response from the terrestrial realm—some changes in diversity of plants, terrestrial vertebrates, and insects, but not enough to really qualify as a mass extinction, nor well enough dated to be demonstrably correlative with the marine extinction. This chapter focuses on the terrestrial events across the Permo-Triassic boundary in an attempt to tease out more clues to the nature of the mass extinction. While the emphasis here is on changes during the Late Permian I have included brief overviews of earlier events to set the stage.

THE TERRESTRIAL VERTEBRATE RECORD

Fish

RECALL FROM the preceding chapter that Pitrat (1973) found a far greater magnitude of extinction in the marine fishes than among the nonmarine fishes. Pitrat's results are quite different from Schaeffer's (1973) and Thompson's (1977), but neither of the latter distinguished between marine and nonmarine fish. The 40 percent extinction among Leonardian nonmarine families compares to an 11 percent extinction in the last stage of the Permian. Pitrat's data are intriguing, for they suggest that the end-Permian extinction was primarily a marine affair and had little impact on freshwater and euryhaline ecosystems. Nonetheless, reliance on the dated compilations of Roemer (1966) suggests a need for a new look at the problem (see Maxwell and Benton 1990).

Tetrapods

The climatic changes of the Permian had a major impact on tetrapods as the amphibians, the dominant vertebrate group on land during the Carboniferous, experienced a steady decline in diversity. Reptilian groups continued to diversify. Long-term accommodation to climatic change appears to have been the driving force behind the replacement of amphibians by reptiles, but vertebrate paleontologists have long recognized four major episodes of nonmarine tetrapod extinction during the Permian and Triassic: at the end of the Early Permian, Late Permian, Early Triassic and Late Triassic. Yet they have been unable to agree on the importance of the Late Permian event because of difficulties assessing the magnitude of the event and correlation problems, which make it unclear whether the marine and terrestrial extinctions occurred at the same time.

Amphibians

Many Carboniferous amphibian groups continued into the Permian only to disappear during the Early Permian. *Eryops* (figure 4.1a) is a typical Early Permian temnospondyl, one of the Carboniferous groups that thrived during the Permian. It reached 2 m in length and had a heavier skeleton than its Carboniferous ancestors. *Eryops* was a semi-aquatic predator feeding on a variety of fish, amphibians, and reptiles

A

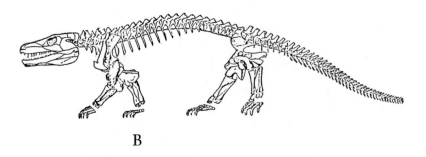

B

Figure 4.1.   Representative Early Permian amphibians. (a) *Eryops*, a temnos-pondyl; and (b) *Seymouria*, a cotylosaur. From Gregory (1952).

and was probably one of the top carnivores during the Early Permian. During the mid-Permian the terrestrial temnospondyls declined, with only the specialized aquatic and fish-eating species surviving (Benton 1990b). Another significant Early Permian amphibian group were seymouriamorph cotylosaurs, like *Seymouria* (figure 4.1b). Closely related to *Seymouria* were the herbivorous Diadectomorpha, including *Diadectes*. The heavy skeleton, short limbs, and the structure of the skull all place this group along the transition between amphibians and reptiles (DiMichele and Hook 1992; Benton 1990b).

Amniotes

The Permian was a time of experimentation and diversification among reptiles, and they may have attained greater morphologic diversity than at any subsequent time in their history. During the Early Permian pelycosaurs were predominant, including *Eothyris*, *Edaphosaurus*, and *Dimetrodon* (figure 4.2a–c; Benton 1990b). By the mid-Permian pelycosaurs reached 150 kg or more in size (DiMichele and Hook 1992). Maxwell (1992) speculates that global warming and increased aridity during the late Early Permian may have done in the pelycosaurs, as four of six families became extinct during the Artinskian and the last two in the following stage. Pelycosaurs were replaced during the mid-Permian by a diverse assemblage of herbivores and carnivores known as the therapsids, which comprise about 80 percent of Late Permian diversity (Colbert 1986). Among the important therapsid groups were the dinocephalians with their fairly massive skeletons and interlocking incisors; both herbivores and carnivores were present. The dicynodonts were an important group of Late Permian herbivores with slender bodies and bird-like beaks enclosing a set of grinding teeth further back in the jaw. Some paleontologists suggest that they may have had complex behavior patterns (King 1992a). Thirty-five genera occur in the Late Permian, but only two, including the widespread genus *Lystrosaurus*, occur in the earliest Triassic. After a mid-Triassic resurgence the clade became extinct at the close of the Triassic (King 1990a). Another clade, the gorgonopsiids, were superficially similar to saber-toothed cats but disappeared with the extinction of the dicynodonts and the dinocephalians, their primary prey, at the close of the Permian. The radiation of these synapsid reptiles sometimes overshadows the many nonsynapsid Permian reptiles. These include the unusual captorhinids with

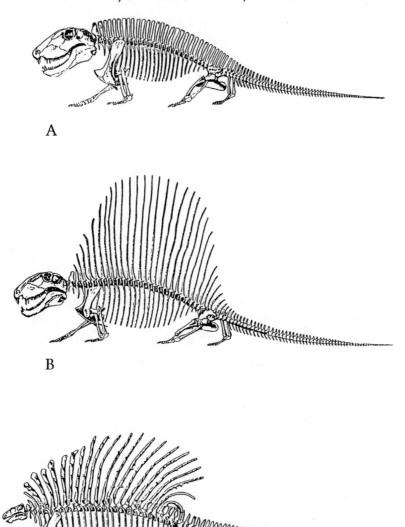

A

B

C

Figure 4.2.   Representative pelycosaurs from the Upper Permian: (a) *Eothyris*; (b) *Dimetrodon*; and (c) *Edaphosaurus*. From Gregory (1952).

their multiple rows of peg-like teeth, pareiasaurs with massive limbs and a huge skull covered with bony knobs and flanges, and the mesosaurs, slender swimmers and the first amniotes to return to the sea (DiMichele and Hook 1992; Benton 1990b).

Extinction Patterns

The first appearance of *Lystrosaurus*, an easily identifiable therapsid, is taken as the beginning of the Triassic in terrestrial sections. This coincides with a marked change in the composition of vertebrate faunas. Earliest Triassic vertebrate faunas have few species but the species that do occur are often abundant, a pattern typical of faunas following an environmental crisis. Unfortunately, the correlations between this terrestrial event and the marine boundary are uncertain. The principal guides to vertebrate extinction patterns across the Permian and Triassic (Pitrat 1973; Olson 1982, 1989; Benton 1985, 1987a, 1988; King 1990a, b; Maxwell 1992) consider amphibians and reptiles together, a practice we will continue here. Although all authors agree that tetrapod diversity dropped in the Late Permian, differences in stratigraphic resolution and taxonomy have produced very different estimates of the magnitude of extinction.

Both Pitrat (1973) and Olson (1982) based their studies on diversity patterns calculated from data in Roemer (1966). Pitrat calculated a 54 percent extinction at the end of the Leonardian followed by a 75 percent drop in the Guadalupian and 63 percent in the Tatarian (figure 4.3). Among the reptilian groups Pitrat recorded, are a total of 28 families from the Tatarian, almost twice as many as from the Guadalupian or Leonardian. However, according to Pitrat, 14 of these 28 families are known only from the Beaufort Group of South Africa and nearby regions, a pattern Pitrat (following Parrington [1948] and Cox [1967]) suggested may represent the unusual preservation of an upland terrestrial fauna in the Beaufort Group. If true, the extinction may represent elimination of a regional endemic fauna rather than a global change in terrestrial ecosystems. Recent sedimentologic studies cast doubt upon the upland nature of the Beaufort Group, however (King 1991; Hiller and Stavrakis 1984).

Olson's (1982, 1989) data are tabulated at a coarser level of resolution than that employed by Pitrat, and suggest an 81 percent extinction among amphibians in the Lower Permian, with diversity dropping from 32 to 8 families and further reductions of 82 percent, 60 percent, and 64 percent during the Middle and Upper Permian and

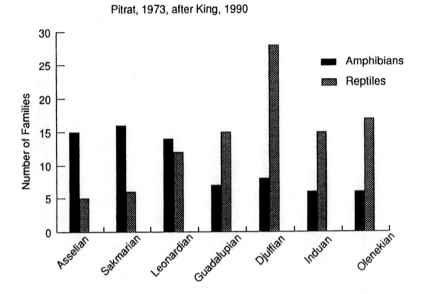

Figure 4.3. Amphibian and reptilian family diversity. From Pitrat (1973), based on the corrected data in King (1990a).

Lower Triassic (figure 4.4), results broadly comparable to Pitrat's. Among the reptiles the therapsids are primarily confined to the Late Permian. In 1989, Olson employed a zonal scheme (although it is not clear that the data are different from his 1982 paper) and found a reduction in the number of vertebrate genera from 90 in the Late Permian to 25 in the earliest Triassic. Olson (1982) argues that a combination of climatic and geographic changes were responsible for the decline at the end of the Permian, but found no evidence for a catastrophic extinction (although, given the low resolution of his data, it is not clear how Olson could have found evidence for a sudden extinction). Like Pitrat, Olson (1989) notes the changes in depositional environment associated with the appearance of *Lystrosaurus*, and suggests that the magnitude of the extinction may be exaggerated by a shift in facies, a hypothesis supported by Hotton's (1967) study of the Beaufort Group.

One potential problem with the data upon which each of these studies is based is the quality of the systematic work, particularly the definition of families. Several more recent papers have employed cladistic approaches to define higher-level taxonomic groups, offering

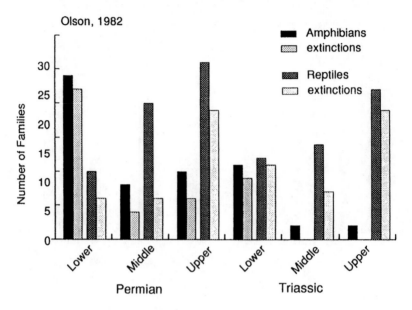

Figure 4.4. Amphibian and reptilian family diversity and number of familial extinctions. From Olson (1982).

what may be a more accurate depiction of diversity trends. Milner's (1990) analysis of temnospondyl amphibians provides important new information on the history of this group. He notes a series of extinctions during the Permian, with a peak during the Kungurian and the disappearance of one family in the Guadalupian and two in the Tatarian. Of greater interest, however, is the indirect evidence he provides for a rapid extinction during the Late Permian. He notes that all post-Permian temnospondyls were apparently of Gondwanan origin, whereas almost all of those disappearing during the Late Permian were exclusively Laurasian, suggesting a regional extinction pattern. Furthermore, the explosive Early Triassic radiation of the group is reminiscent of a postextinction rebound. While Milner admits that the depauperate nature of Late Permian amphibian faunas is in part a consequence of poor preservation, he suggests that the rapid radiation in the Early Triassic is at least suggestive evidence of a relatively rapid extinction.

Benton (1985, 1987a) also developed a new data set with systematic assignments revised on the basis of cladistic analyses. He identified the extinction of 8 families (58 percent) of nonmarine tetrapods during the Sakmarian-Artinskian (largely of amphibians and pelycosaurs) and the disappearance of 21 families (49 percent) during the Tatarian. According to Benton's analysis, 15 of the Tatarian families belonged to the therapsids and spanned all size ranges. Unlike Pitrat, Benton could not find any geographic control on the extinction patterns. During the Scythian another extinction occurred, removing 7 families, but because of numerous originations overall diversity did not decline.

Recently Maxwell (1992) reanalyzed Benton's data set. He found a steady decrease in amphibian diversity through the Lower Permian from 23 families to 9 after the Artinskian, with 7 of 15 (46 percent) disappearing during the Artinskian (figure 4.5). In contrast to Milner's results, Maxwell found relatively steady amphibian familial diversity through the remainder of the Permian and into the Triassic, although increases in extinction frequency occurred during the Capitanian, Tatarian, and Scythian. Reptiles experienced extinction peaks during the same intervals, with a particularly extensive extinction during the Tatarian (21 of 27 families, or 78 percent). Interestingly, all but one of the 9 tetrapod families that survived the extinction became

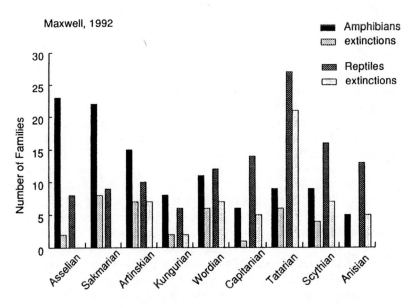

Figure 4.5. Amphibian and reptilian family diversity and number of familial extinctions. From Maxwell (1992), based on data from Benton (1987c).

extinct during the Scythian. The net effect of the decline in amphibians and the diversification of the therapsids was the progressive replacement of amphibians by nonamniote tetrapods as the dominant members of terrestrial communities. Maxwell puts an interesting twist on the difference in extinction numbers between amphibians and reptiles. Rejecting the climatic deterioration hypothesis advanced by Pitrat (1973), Maxwell argues that the difference simply reflects diversity. More amphibian families in the Early Permian led to a greater number of extinctions among amphibians, fewer families during the Late Permian led to fewer extinctions.

Finally, Sloan (1985) described ten major Permian extinctions, two during the Artinskian/Kungurian, two in the Wordian/Capitanian, and six extinctions during the last stage of the Permian. He found a preferential removal of the largest-sized animals and those with the fewest mammalian characters, resulting in the progressive increase in the mammalian nature of the fauna. He suggests that over 50 percent of the morphologic change from the earliest reptiles to the earliest mammals occurred at this time. Unfortunately, Sloan has not published his findings, so the quality of his data and the accuracy of his conclusions cannot be determined.

Each of these analyses identified an episode of apparently increased extinction during the Tatarian, but many authors have raised the troubling possibility of preservational bias (Colbert 1973; Padian and Clemens 1985; Pitrat 1973; Olson 1982, 1989; Maxwell and Benton 1990; Maxwell 1992). The problem is essentially that given the complex pattern of ecological and evolutionary change during the Permian and the spotty distribution of well-preserved vertebrate faunas in the latest Permian (Olson 1989), how does one distinguish a mass extinction from an ecological shift (Padian and Clemens 1985)?

Detailed studies within a single region with many well-exposed localities is one solution, and fortunately the Karoo Basin of South Africa fits the bill. The Karoo contains many of the most complete terrestrial Permo-Triassic sections (Hotton 1967; Kitching 1977; Keyser and Smith 1977). King (1990a) analyzed generic diversity patterns through the Beaufort Group based on Kitching's (1977) five biostratigraphic zones for the Karoo, which range in age from the Guadalupian to the Middle Triassic. She found a gradual decline in generic diversity through the boundary from a peak of 80 genera in the *Daptocephalus* Zone to 18 in the *Lystrosaurus* Zone in the earliest Triassic, an overall extinction of 78 percent, over perhaps 8 million years (figure 4.6). More recently, King (1991) evaluated the effect of taxo-

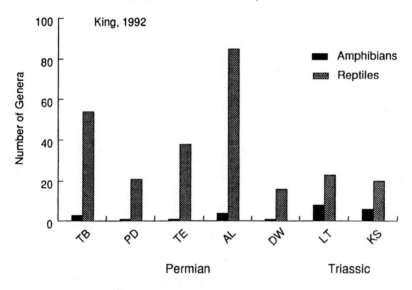

Figure 4.6. Amphibian and reptilian family diversity in South Africa based on Kitching's (1977) Assemblage Zones. The Permo-Triassic boundary lies at the base of the *Lystrosaurus-Thrinaxodon* Assemblage Zone. Key to Assemblage Zones: TB = *Tapiocephalus-Bradysaurus*; PD = *Pristerognathus-Diictodon*; TE = *Tropidostoma-Endothiodon*; AL = *Aulacephalodon-Cistecephalus*; DW = *Dicynodon lacerticeps-Whaitsia*; LT = *Lystrosaurus-Thrinaxodon*; and KS = *Kannemeyeria simmocephalus*. From King (1991).

nomic revision, enhanced biostratigraphic resolution, and sampling bias on diversity patterns within the Karoo. The influence of the Signor-Lipps effect (discussed in chapter 3) was of particular interest. After correcting for these factors, King identified two episodes of increased extinction, the first during the *Tapinocephalus-Bradysaurus* Assemblage Zone (approximately Capitanian in age), and the *Aulacephalodon-Cistecephalus* Assemblage Zone (Tatarian, but not at the Permo-Triassic boundary). The ability to resolve the single peak found in earlier studies into two peaks was a result of a finer time scale rather than any other correction. Moreover the results indicate that the Signor-Lipps effect was not responsible for the apparently prolonged extinction, rather the pattern may be real.

In summary, the end-Permian was an interval of enhanced extinction among terrestrial tetrapods, apparently including temnospondyl amphibians, but there is no evidence for a catastrophic extinction.

Although the available data remain poor and subject to a variety of possible biases, they suggest a continuing decline from the middle of the Late Permian into the Triassic. The end-Permian terrestrial episode was the largest of at least four or five mass extinctions during the Permian and Triassic, involving a 67 percent decline among amphibians and a 78 percent drop among "reptiles."

THE PALEOPHYTIC-MESOPHYTIC
FLORAL TRANSITION

THE PERMIAN was a time of reorganization of plant communities, much like the transition from the Paleozoic Evolutionary Fauna to the Modern Fauna in the marine realm. This reorganization marks the end of the Paleophytic Flora and the rise of the Mesophytic Flora (figure 4.7). The Paleophytic Flora was dominated by pteridophytes and primitive gymnosperms, which were replaced by the more advanced gymnosperms of the Mesophytic Flora. Arborescent (tree-like) sphenopsids and cordiate gymnosperms disappeared during the

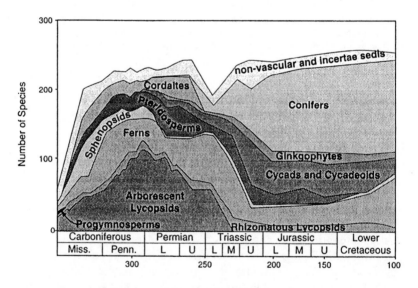

Figure 4.7. Changing patterns of global plant diversity during the Permian and Triassic, illustrating the decline of the Paleophytic Flora and the rise of the Mesophytic Flora. Adapted from Niklas et al. (1985).

126

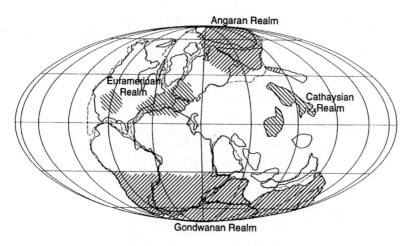

Figure 4.8. Major floral biogeographic realms (similar to provinces). Based on the work of Ziegler (1990).

Permian and the arborescent lycopsids declined dramatically in number. All were important components of the Paleophytic Flora. At the same time cycads, cycadeoids, ginkgophytes, peltosperms, and, most importantly, the conifers began to diversify, establishing the Mesophytic flora.

The plants of the Paleophytic are divided into more floral realms or provinces than any period until the Late Tertiary (figure 4.8) (Ziegler 1990; Chaloner and Meyen 1973). Plant communities in the Angaran province of the Siberian Platform and adjacent regions were dominated by a distinctive assemblage of cordiates. In the higher latitudes of Gondwana glossopterid pteridosperms were prominent. By the Late Permian the Euramerican province of the Late Carboniferous and Early Permian had divided into two provinces, one centered in the southwestern United States, the other in western Europe. In the latter, broad-leaved pteridosperms were the most important group along with cordiates, pecopterid ferns, and other pteridophytes. Similar taxa are found in the distinctive floras of the Cathaysian Province, centered in North and South China. This assemblage is dominated by *Gigantopteris* and its allies (Ziegler 1990; Chaloner and Meyen 1973; Meyen 1982). Coal balls demonstrate the persistence of Late Carboniferous floras of lycopsids and cordiates until the end of the Permian in South China (DiMichele 1992).

In the past paleobotanists associated the Paleophytic-Mesophytic transition with the Permo-Triassic boundary but more refined studies have traced the origin of the Mesophytic flora to the Late Carboniferous (DiMichele and Aronson 1992). The transition between the two floras began in low latitudes during the Lower Permian (Artinskian) and spread toward the poles, with the final transitions occurring in the Early Triassic. Mesophytic conifers apparently evolved in dry upland habitats during the Late Carboniferous and began to spread out of the uplands into the lowlands, first in Euramerica and then North China, as the lowland areas began to dry out following the Permo-Carboniferous glaciation. In the process dryland floras replaced a variety of ferns and pteridosperm groups (Frederiksen 1972; DiMichele and Aronson 1992). Similarly, the extinction of the Paleophytic ferns, lycopsids, and sphenopsids during the Permian reflects the disappearance of the wet habitats in which these groups predominated. Within a single region the transition from Paleophytic floras to Mesophytic floras may be as short as 5 million years but during the transition one finds a mix of floras, some of Paleophytic aspect, others presaging the Mesophytic Era (figure 4.9) (Meyen 1973; Knoll 1984; DiMichele and Hook 1992). The pattern of transition has been

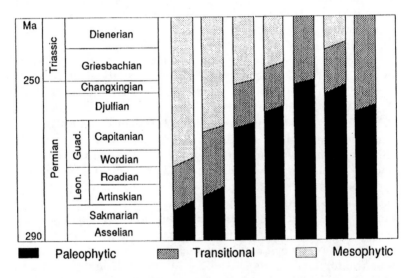

Figure 4.9. The Paleophytic-Mesophytic transition; the columns record the transition within single regions: (a) North America; (b) Western Europe; (c) Russian Platform; (d) Northern Eurasia; (e) Northern Cathaysia; (f) Southern Cathaysia; (g) Australia. After Knoll (1984).

128

studied intensively in Europe and North America by DiMichele and Aronson (1992), who find that plants of Mesophytic aspect first appeared in the fossil record in dry to seasonally dry habitats. As they moved into lowland environments they occupied a different set of habitats from the older, Paleophytic floras, so the floras remained in distinct habitats within a single region. Plants adapted to seasonally dry conditions expanded and replaced older plants in all lowland habitats as climates became warmer and drier. Knoll (1984) noted the many similarities between this transition and Neogene changes in plant diversity caused by a combination of climatic and tectonic changes.

Total family diversity dropped by about 50 percent during the 25 million years over which the transition occurred (from 30 families in the Lower Permian to 15 in the Middle Triassic) (Knoll 1984).

Plants are found not only as impressions of leaves and other plant parts, but also as pollen and spores. The fossil record of pollen and spores is very responsive to local and regional changes in the abundance of plants and thus faithfully preserves a record of environmental changes and other events that affect plant abundance. The Paleophytic/Mesophytic transition is marked in the pollen record by the appearance of taeniate pollen in a wide variety of groups. The pleated structure in the middle of a taeniate pollen grain may represent an adaptation to increased aridity (Traverse 1988).

At the boundary itself, changes in the pollen record indicate rapid climatic change. In the southern Alps the abundance of fungal remains and other terrestrially derived plant remains increases abruptly (see also chapter 2). At the same time many characteristic Late Permian gymnosperm pollen types disappear from the record. Interestingly, this coincides with the elimination of the last Permian marine fossils in the lower portion of the Tesero Horizon (Visscher and Brugman 1988). A similar change is found across the Permo-Triassic boundary in drillholes in Israel, where the boundary is marked by a red clay and a change in pollen types (Eshet 1990, 1992). At some drillholes virtually all of the Permian pollen types disappear and are replaced by a different suite of pollen. The continuity of these sections is uncertain and an unconformity seems likely, but the general pattern is one of increase in terrestrially derived clastic material, reworked organic material, and fungal spores, as well as a shift in pollen assemblages. Most of the Triassic forms are acritarchs, organic-walled marine microfossils of uncertain affinities but representing an increase in the abundance of phytoplankton. The increase in acritarchs during the earliest Triassic probably represents an opportunistic expansion of this group following the extinction (Eshet 1992).

The fungal spike at the Permo-Triassic boundary is intriguing, since fungi require moisture and could not germinate in a highly arid climate. If these sections are relatively continuous, they may record either a mass dying producing abundant decaying biomass, or the relatively sudden extinction of many elements of the terrestrial Permian flora and the expansion of fungi (Visscher and Brugman 1988). Alternatively, the fungal spike may be a taphonomic artifact, representing the differential preservation of fungal spores (protected by their outer coating of chitin) and the removal of other, less resistant spores and pollen (Mamay and Wing 1992). The relative increase in abundance could also represent a condensed interval caused by a decline in the sedimentation rate and consequent concentration of fungal spores. Which of these possibilities is more likely will require further study. Past studies have been concentrated immediately around the putative boundary, so we have no information on long-term changes in fungal diversity over a longer interval during the Late Permian.

The pollen record of western Australia suggests the earliest Triassic transgression may be responsible for at least some of the abrupt changes. Diverse plant communities assigned to the *Dulhuntyispora* assemblage disappear at the boundary and are replaced by an impoverished assemblage during the earliest Triassic. Similar changes are associated with other rapid transgressions in the mid-Carboniferous, the Late Triassic, and near the end of the Early Cretaceous, suggesting that the elimination of this diverse Permian plant assemblage may have been due to flooding of broad coastal lowlands during the rapid transgression (Balme and Helby 1973), a model that may well explain many other terrestrial events.

The general picture one receives from the plant record is at best equivocal. Certainly the decline of the Paleophytic Flora and the rise of the Mesophytic Flora was a prolonged event related to the origin of new groups like the conifers and the opportunities presented by the end of the Permo-Carboniferous glaciation. At the same time, the pollen record does indicate an ecological disturbance on land at the close of the Permian and an apparent increase in terrestrial input into marine sections. This suggests that a brief environmental disturbance may have occurred in western Tethys near the Permo-Triassic boundary. While this pattern remains intriguing, and clearly deserves further study, to date the record does not provide any convincing support for a catastrophic terrestrial extinction.

INSECTS DURING THE PERMIAN

LIKE MANY invertebrate paleontologists I have long held the prejudice that insects get ground to a pulp soon after dying and thus have a negligible fossil record. In fact, insects have a wonderful fossil record and the Permian has one of the most diverse and best-known insect faunas prior to the Tertiary (Labandeira and Beall 1990; Wootton 1990; Shear and Kukalova-Peck 1990; Carpenter and Burnham 1985). The fascinating geologic history of insects reveals the Permian as a time of transition as the palaeopterous insects were replaced by the neopterous insects. Almost all palaeopterous insects have a simple hinge attaching the wing to the body, so the wings cannot be folded back over the body at rest (figure 4.10). Today this group is limited to the modern dragonflies and mayflies. Neopterous insects include virtually all modern groups (about 98 percent). Their ability to fold their wings allowed them to move into many new habitats (Carpenter and Burnham 1985). In addition, most Neopterous insects are also holometabolous, meaning they undergo complete metamorphosis with a resting stage—or pupa—separating the ecologically and morphologically distinct larval stage from the adult stage.

Both paleopterous and neopterous orders appeared by the Late Carboniferous, but the Permian was a time of rapid diversification. Twenty-two of the 36 described insect orders are known from the Permian. The number of orders that disappeared during the Late Permian varies, according to different authors, because of taxonomic problems. Sepkoski and Hulver (1985) and Carpenter and Burnham (1985) each list eight orders, although only five are the same. Conrad Labandeira has looked into this problem and recognizes the disappearance of nine orders during the Late Permian. An additional ten suffered severe drops in diversity (Sepkoski and Hulver 1985). Although Triassic insect faunas are spotty, a diverse Scythian fauna restricts the age of this extinction to the Late Permian (Sepkoski 1992b) but we cannot further constrain the timing. By way of comparison, only one or two insect orders have became extinct since the Permian.

Many of these extinctions occurred among the palaeodictyopteroids, a diverse assemblage of four or five orders (depending on who is counting) with mouth parts adapted for piercing and sucking, probably largely on sap (similar to the modern order Hemiptera), although the smaller forms may have been bloodsuckers. Shear and Kukalova-

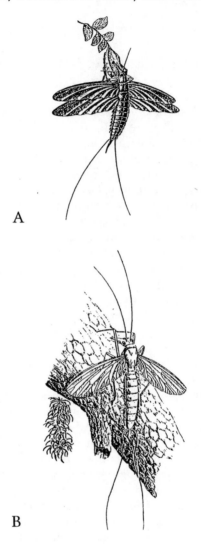

A

B

Figure 4.10. Representative Palaeodictyopteroids from the Permian. Note the nonfolding wings. Both insects are shown piercing and sucking, but it is unclear whether they actually followed this feeding strategy. (a) *Perothemistis* sp., (b) *Goldbergia*. From Rohdendorf and Rasnitsyn (1980).

Peck (1990) raise the possibility that some groups may have fed on insects or even tetrapods. The palaeodictyopteroids began to decline during the mid-Permian and disappeared entirely by the end of the period, but even those orders that cross the boundary show some changes. For example, the Paleozoic dragonflies and mayflies are placed in different suborders from their post-Paleozoic descendants (Wootton 1990).

Our understanding of Permian insect evolution and extinction patterns remains in its infancy, but the close association between plants and insects had clearly begun by the Permian (Shear and Kukalova-Peck 1990; Labandeira and Beall 1990; Scott, Stephenson, and Chaloner 1992). This might suggest that the transition in insect assemblages was related to the Paleophytic-Mesophytic floral transition. Certainly the reduction in the palaeodicyopteroids during the mid-Permian indicates that this group may have declined because of the disappearance of their host plants. However, the major drop in insect diversity occurred in the Late Permian and seems difficult to relate directly to the climatically driven changes in plant floras. Labandeira (1992) has pointed out that holometabolous insects may have been favored as climatic seasonality increased during the Permian because they can enter a resting phase during the inclement season. This adaptation may have relegated the Palaeoptera and the nonholometabolous Neoptera to the fringes of insect evolution, yet these changes appear to have largely predated the end of the Permian. Clearly there is considerable room for further work on the relationship between the mass extinction and insect diversity patterns.

As we turn from the patterns of disappearance to the causes of the end-Permian mass extinction we find evidence of some correlative terrestrial extinctions. The evidence from tetrapods is at best equivocal, with some specialists arguing for a major extinction peak during the Late Permian, while others emphasize the poor quality of the data and suggest that the increase in extinction was somewhat less. The Late Permian floral record provides no evidence for such an extinction. On the other hand, the discontinuity in the pollen record and the spike in fungal spores at the boundary do indicate a short-lived environmental disturbance. In general the pattern of tetrapod and plant diversity during the Permian appears to be best explained as a response to long-term, gradual climatic changes associated with the end of the Permo-Carboniferous glaciation and the subsequent onset of global warming.

V

Pangea and All That

THE PERMO-TRIASSIC boundary marks a major transformation in geologic patterns as the formation of the supercontinent Pangea and resulting reorganization in mantle convection led to changes in the style of tectonic processes, orogenic pattern, mantle heat flow, paleomagnetic reversals, climate, plutonism, and sedimentation (Holser and Magaritz 1987). To a geologist, these differences distinguish the Paleozoic from the Mesozoic-Cenozoic just as surely as the change in fossils. It may seem obvious that such extensive change in geologic process was related to the mass extinction at the end of the Permian, but the causal connection remains unclear. Scientists usually demand more of an explanation than the coincidence or correlation of two events in time, but some geologists have proposed a variety of mechanisms to provide a causal connection between the geologic events and the extinction.

Evaluating these hypotheses requires a detailed knowledge of the geologic, geophysical, geochemical, and climatic framework of the Permian, and an understanding of the changes between the Paleozoic and Mesozoic. These topics are the subject of the next four chapters. This chapter covers the geologic and geophysical changes between the Paleozoic and Mesozoic, beginning with the formation of Pangea, changes in sea level, and the various volcanic episodes of the Upper Permian. The next two chapters focus on Permian climates and on geochemical changes, respectively. Then in chapter 8 we return to the implications of this discontinuity in geologic processes for the end-Permian mass extinction.

THE FORMATION OF PANGEA

AS EARLY as 1885 Melchinor Neumayr, a Viennese stratigrapher and son-in-law of the famous geologist Edward Suess, described a Jurassic seaway stretching from southeast Asia to the Caribbean. Suess developed the concept further, particularly in his influential book *Das Antlitz der Erde* [*The Face of the Earth*] (1901) in which he identified a suite of tropical to subtropical fossils deposited in a broad seaway between the northern and southern continents. Suess claimed that this Tethyan seaway developed by the Late Paleozoic and persisted into the Tertiary (Tollmann and Kristan-Tollmann 1985). Considerable debate ensued between "fixists," who viewed Tethys as a broad geosyncline that became progressively smaller during the Phanerozoic, and "mobilists," who argued for a narrower belt between drifting continents (Şengör 1987).

This debate, and a related one over the supercontinent of Pangea, was ultimately resolved by the discovery of plate tectonics and continental drift, whose outline should be familiar to many: during the late Paleozoic the southern supercontinent of Gondwana collided with the northern supercontinent of Laurasia, creating a series of orogenic belts and forming the supercontinent of Pangea. Pangea began to break apart during the Triassic, with dispersal accelerating in the Jurassic and Cretaceous and continuing even today.

Reconstructing the movement of the various bits of Pangea is fairly accurate for as far back as the Jurassic because of evidence supplied by magnetized stripes on the seafloor. For the pre-Jurassic the lack of preserved oceanic crust means that other, inherently less reliable techniques must be used. Pangean reconstructions (figures 5.1 a–d) rely on evidence of structural similarities between continents, on the

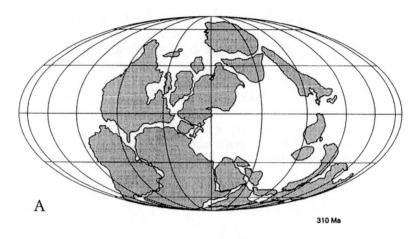

A

310 Ma

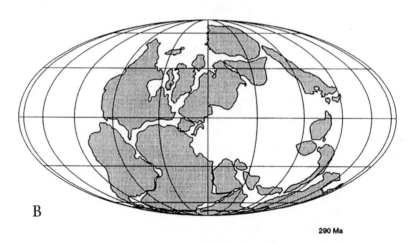

B

290 Ma

Figure 5.1.   Pangean reconstructions from the computer program *Terra Mobilis*, written by Chris Scotese, showing likely continental configurations from the Early Permian into the Triassic. The placement of terranes in the Tethyan region is poorly constrained. (A) 390 million years ago; (B) 290 million years ago; (C) 270 million years ago; (D) 250 million years ago. The output from *Terra Mobilis* has been modified, with North and South China moved further to the west, as seems more likely.

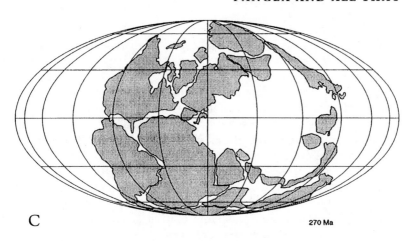

C   270 Ma

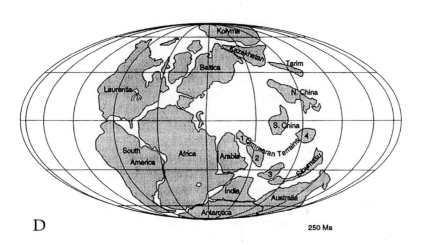

D   250 Ma

distribution of both terrestrial and marine fossils, on paleomagnetic evidence, and on paleoclimatic indicators, such as the distribution of coals or glacial sediments.

Since the same data can oftentimes be interpreted in various ways and different types of data may conflict, disagreement persists over many details of these reconstructions, even for the positions of Laurasia and Gondwana (Lottes and Rowley 1990). These disagreements reflect differences over the most accurate poles of rotation, the extent of continental deformation, the position of the paleomagnetic pole, and apparent polar wander paths (see Lottes and Rowley 1990; Scotese and McKerrow 1990). Furthermore, since paleomagnetic evidence provides only paleolatitudes, the longitudinal positioning of the continents depends largely on fossil information, and because of different interpretations, introduces substantial variation between reconstructions. Thus, these paleogeographic reconstructions must be treated as approximations, with the accuracy of the reconstruction varying in time and space. The following discussion of these events is based on important summary papers from a recent symposium on Paleozoic palaeogeography and biogeography (McKerrow and Scotese 1990), particularly the work of Scotese and McKerrow (1990), Lottes and Rowley (1990), Nie, Rowley, and Ziegler (1990), and Ziegler (1990). Other sources include Smith (1988), papers in Nakazawa and Dickins (1985) and in Audley-Charles and Hallam (1988), particularly those of Metcalfe (1988) and Şengör et al. (1988).

An early reconstruction of Pangea (Wilson 1966) included a large, triangular-shaped Tethyan ocean between Gondwana and the eastern portion of the Laurasian supercontinent (known as the Baltic-Kazakhstan-Siberian cratons). Other early workers in plate tectonics related the destruction of Tethys to the formation of the Alpine-Himalayan mountain belt with Tibet, China, and most of the Middle East included in the Laurasian supercontinent. An immediate practical difficulty with these reconstructions soon arose, for the closure of such a broad Tethyan ocean would have left considerable geologic evidence, such as ophiolite sequences and evidence of subduction zone volcanism. However, the evidence simply was not there. Never at a loss, geologists quickly proposed a variety of alternative reconstructions.

In the most generally accepted of these reconstructions, known as Pangea A (Hallam 1983), the formation of Pangea began with the collision of Gondwana and Laurentia (North America, Greenland, and Baltica [northern Europe west of the Ural Mountains]) in the Early

Carboniferous. Laurentia and Gondwana had been adjacent since at least the Devonian but deformation did not begin until the Early Carboniferous. As the collision progressed, Gondwana rotated clockwise relative to Europe with deformation beginning in the northeast along the southern margin of Europe and progressing southwest along the Variscan fold belt of northwest Europe and the Appalachians. Eventually the orogenic activity reached the Ouachita Mountains of Oklahoma in the Late Carboniferous and Early Permian. To the east, Kazakhstan collided with the southwest margin of Siberia in the Middle Carboniferous and this combined block rotated clockwise, closing the Uralian Seaway. The Uralian Orogeny began as this block collided with the northeastern margin of Baltica in the Early Permian (Scotese and McKerrow 1990; Lottes and Rowley 1990). This reconstruction places South America against the southeastern margin of North America, Africa against the southern margin of the eastern seaboard of North America, and Baltica somewhat further north (Scotese 1984; Van der Voo and French 1974; Lefort and Van der Voo 1981; Van der Voo, Peinado, and Scotese 1984).

By the mid-1980s several alternate reconstructions of Pangea were proposed, reflecting uncertainties in the available geologic and paleomagnetic information. Pangea B aligns northwest South America with eastern North America, rotating Gondwana by 35° (Morel and Irving 1981; Irving 1977). Lin, Fuller, and Zhang (1985) suggested that Pangea A was a short-lived intermediate and proposed a variant of Pangea B that again placed South America against eastern North America, but positioned Baltica against the northern margin of North America. The Lin and Fuller reconstruction also has a small Tethyan ocean, partly occluded by the equatorial placement of the North and South China blocks.

The third reconstruction, Pangea C, never garnered much support. In this model North America is connected to Baltica via Newfoundland and Greenland, but not to Europe. South America is displaced far to the east relative to the other reconstructions and abuts the southern margin of Baltica. This reconstruction fails to account for the collisionary tectonics along the Appalachian-Ouchita front and is at variance with considerable other geologic evidence. Moreover, it requires some 6,000 km of right-lateral shear between Laurasia and Gondwana between the mid-Permian and mid-Triassic (Hallam 1983). The 15–30 cm/year of movement along an unidentified strike-slip zone between Laurasia and Gondwana required by this shear zone seems a bit unlikely (Hallam 1983; Smith, Hurley, and Briden 1981).

139

Detailed paleomagnetic, paleobotanical, and geologic studies in the late 1980s provided increasing support for the broad outlines of the Pangea A reconstruction. As always, the devil is in the details, and the patterns of continental accretion and dispersal are particularly complex in Asia. Perhaps the most important conclusion from these studies is the recognition that Pangea initially formed by the end of the Early Permian, then reformed toward the end of the Triassic as a melange of tectonic blocks began to form present-day Asia.

Today the Eurasian, Pacific, and Indo-Australian plates interact with a swarm of minor plates in the Circum-Pacific and Alpine-Himalayan tectonic belts. Because this region also contains many of the best Permo-Triassic boundary sections, reconstructing Permian geography is critical to understanding biogeographic aspects of the extinction. For example, some reconstructions propose a large number of independent tectonic units, and potentially independent biogeographic units; of course the two are *not* synonymous, although they are frequently confused. These isolated realms could have served as refugia during the mass extinction. The impact of such refugia would be enhanced if the causes of the extinction were largely confined to Pangea.

Asia was produced by the collision of a series of small microplates, or terranes, and the sutures between terranes are demarcated by ophiolite suites, fault zones, and both fossil and geologic discontinuities. Postcollisional tectonics makes reconstructing the original geography of the terranes difficult, and even the number of independent tectonic terranes is uncertain (Nie, Rowley, and Ziegler 1990; Şengör 1987). Suture zones extending from Turkey through China and southeast Asia enclose two different clusters of terranes: the Cathaysian terranes of North China (Sino-Korea), South China (Yangtze), Eastern Qiangtang, Tarim, and Indochina; and the Cimmeran (Şengör 1987; Şengör et al. 1988), or peri-Gondwanan (Nie, Rowley, and Ziegler 1990), terranes of the Helmand, central Iran, western Qiangtang, Lhasa, and Sibumasu (Shan Thai and Malaya) blocks that are presently distributed from modern-day Turkey and Iran, through Tibet, into Malaysia.

Three models have been advanced for the distribution of these terranes prior to accretion (Smith 1988): first, that oceanic Tethys never existed because the earth's diameter was 20 percent smaller in the Permian and has expanded to its present size over the past 250 million years; second, that the terranes were scattered in no particular order across the Tethyan ocean; and third, that the terranes formed a coherent unit (generally named Cimmeria). Thus one or more of the Asian sutures corresponds to a former ocean basin that closed during the Permian and Triassic, moving the Cimmerian continent north across the ocean basin and into Laurasia.

My personal favorite is the expanding earth hypothesis — not because I believe it, which I do not—simply because the consequences would be the most amusing. Proponents of the expanding earth hypothesis contend that there is less overlap and a closer fit between adjacent continents if an earth with a diameter of 80 percent of the present diameter is used to reconstruct Pangea (Owen 1983; Carey 1983). These models usually assume a linear expansion rate over at least the past 200 million years. One can in fact advance plausible theories of the earth in which the expansion of the mantle is fueled by mineral phase changes, while vertical tectonics drives surficial geology. Unfortunately, neither amusement value nor theoretical plausibility is much of a reason to believe in an expanding earth in the face of overwhelming empirical evidence to the contrary. Hallam (1984a) provides an excellent critical review of various expanding-earth scenarios. Therefore, excluding this possibility, we are left with the other two hypotheses.

There are many similarities between the sequential addition hypothesis and the archipelago hypothesis and it is important not to overstate their differences. Both envision the distribution of the Cathaysian terranes in tropical to subtropical latitudes and their sequential accretion to form Asia during Permian and early Mesozoic times (Nie, Rowley, and Ziegler 1990). For example, the South China Block collided with the North China Block in the Late Permian along the Qinling suture; rotation between the blocks continued into the Triassic (Watson et al. 1987). The archipelago model postulates a long, arcuate Cimmerian continent that detached from northern Gondwana near the end of the Early Permian, moved north, and collided with Asia (Şengör 1987; Şengör et al. 1988) after the collision of the dispersed Cathaysian blocks. Most of the collisional events occurred in the Mesozoic and Cenozoic as the northward movement of Cimmeria closed out Paleo-Tethys and opened Neo-Tethys in its wake. This division of Tethys into Paleo-Tethys and Neo-Tethys is the primary difference between the two models. Recent paleomagnetic results (Huang et al. 1992) indicate that the South China Block and the Qiangtang Terrane (now part of Tibet) were in close proximity in equatorial latitudes during the Late Permian. These results also confirm that these two blocks were isolated from the Lhasa and the

Shan-Thai-Malay terranes to the south, and from Tarim and the North China Block to the north.

Evaluating models based largely on geologic evidence is best done using fossil data, particularly from terrestrial plants and marine invertebrates. Plants provide information on paleolatitude and climate, and floral similarities may indicate proximity. Seed plants in common between different regions are of particular importance since they could not disperse through salt water. For example, the distribution of *Gigantopteris* throughout the Cathaysian microcontinents establishes that these tectonically distinct terranes maintained a geographic connection that allowed dispersal of *Gigantopteris* seeds, although the nature of this connection is unclear (Nie, Rowley, and Ziegler 1990; Ziegler 1990).

The Cathaysian floral realm dominated by *Gigantopteris* is one of four major floral realms identified in the Late Paleozoic (figure 4.8). These realms are: the south temperate Gondwanan Realm in Gondwana, dominated by *Glossopteris*; the north temperate Angaran Realm in Siberia and Kazakhstan; the Euramerican Realm; and the warm temperate to tropical Cathaysian Realm extending through northern Gondwana and the east Asian (Cathaysian) microcontinents. Difficulties in agreeing on the affiliations of floras in northern Gondwana, Saudi Arabia, and Turkey have produced considerable debate about the limits of the Cathaysian flora (Ziegler 1990). Ziegler emphasizes climatic controls in establishing floral similarity rather than a more rigorous biogeographic approach; such an approach may confuse morphologic convergence driven by climatic necessity with similarity due to evolutionary descent. Nonetheless, the affinities between the east Asian and Gondwanan elements of the Cathaysian flora seem well-substantiated. The changing boundaries of these floral realms are important clues to the evolving tectonic and biogeographic borders. Floral evidence supports connections between the Cathaysian terranes, but gives only equivocal data on the existence of the Cimmerian continent.

Marine invertebrates are another important source of biogeographic information. Using statistical techniques that assess the similarities between biogeographic units on the basis of shared endemic (or localized) rugose coral genera in East Asia, Smith (1988) demonstrated that the Cathaysian corals, like the plants, comprised a single, tropical biogeographic realm that was isolated from central Asia until the Late Permian. Moreover, no sudden breaks or discontinuities in the biogeography of corals are evident, suggesting that their distribution was controlled by latitudinal climatic gradients. Smith's results provide no support for Şengör's Cimmerian continent, nor for a difference between Paleo-Tethys and Neo-Tethys.

Applying fossil evidence to test tectonic reconstructions is not always as straightforward as it may appear from these examples. Differences in climate, available environments, soil type, or other variables may allow dissimilar plants and animals to exist on the same tectonic block while the faunas or floras are interpreted as belonging to disjunct biogeographic units. Likewise, superficially similar plants and animals may evolve under similar environmental conditions on different continents. For example, the distributional data for *Gigantopteris* is confounded by frequent misidentifications, a common problem that demonstrates the importance of good systematics. Smith's (1988) analysis of rugose corals demonstrates that the strong latitudinal temperature gradients typical of the Permian can generate dissimilar faunas that may be incorrectly interpreted as belonging to different biogeographic units. Finally, and most important, biogeographic analysis identifies biogeographic units that do not necessarily correspond to tectonic units. As pointed out earlier, the Cathaysian terranes appear to belong to a single biogeographic unit, but paleomagnetic evidence indicates that they form several tectonic units. This contrast between biogeographic and tectonic units has been overlooked by some geologists working on plate reconstructions, but obviously the difference between these two units limits the reliability of biogeographic data in testing continental reconstructions.

How well does the fossil evidence match the sequential addition and archipelago models for Tethys? It would seem that the paleontologic evidence, the abundance of sutures in China and their lack of extensive ophiolites, and the multitude of tectonic blocks support the sequential addition model. Thus, numerous tectonic terranes, some as large as the South China block, were likely distributed through Tethys in Late Permian times. These blocks could have provided refugia if this region was less affected by the extinction than elsewhere, as is implied by mechanisms closely tied to Pangea. If the extinction decimated the fauna on these tectonic blocks as well, we must seek extinction mechanisms with a more global impact.

In summary, the formation of Pangea was largely complete by the end of the Early Permian, well before the onset of the mass extinction. The North China-Tarim block collided with Siberia near the end of the Permian, but the separate tectonic blocks in the Tethyan ocean include some of those with excellent marine records of the latest Permian.

PHYSICAL EFFECTS OF PANGEA

CONTINENTAL DISTRIBUTIONS have an obvious impact on the distributions of plants and animals, but also modify climate patterns, atmospheric and oceanic circulation, and, evidently, even the internal dynamics of the earth. These biologic influences will be discussed in chapters 6 and 7, and the climatic effects in the next chapter. Here we will concentrate on the physical consequences.

Changes in the Earth's Geoid

The earth's geoid is the hypothetical surface of the earth corresponding to mean sea level, and departures from the geoid reflect deviations in the mass distribution of the earth. In other words, imagine the surface of a hypothetical smooth earth with the same mantle structure as our own. Variations in the distribution of mass within the mantle produce broad, irregular uplifts and shallow dimples; such positive and negative anomalies are more difficult to identify because of the irregular surface of our earth. Large amounts of mantle-derived rock form positive geoid anomalies near subduction zones, hot spots, and zones of massive Cretaceous volcanism (Chase 1979). Curiously, both positive and negative long-wavelength geoidal anomalies correspond more closely to the former positions of the continents than to present tectonic features, suggesting a lasting effect in the mantle.

Geologists have produced several explanations for this association between geoidal anomalies and past continental position. Among the most interesting is that of Anderson (1982), who suggested that alternations between supercontinental accretion and breakup reflect changing patterns of mantle dynamics. If the insulating effect of Pangea reduced heat flow from the mantle, thermal expansion and partial melting of the mantle and lower crust should result. Eventually this thermal expansion would lead to the breakup of Pangea. Moreover, since continental thickness increases linearly with continental size, the insulating effect of a supercontinent is greater than would be expected just from the increase in area alone (see below). Anderson's hypothesized relationship between mantle convection, the geoid, and cycles of supercontinental accretion and dispersal has been largely confirmed by both observational data and simulation studies (Gurnis 1988). Pangea was located above the current Pacific and Atlantic-African geoidal highs (Chase 1979) and the major continents have been drifting toward geoidal lows since it began to break up (Olson, Silver, and Carlson 1990). In addition, many present-day hotspots occur on the margin of the geoidal highs and first developed in the early Mesozoic. Unlike Anderson, Chase (1985) suggests that the thermal uplift of Pangea may be a consequence of subduction zones around the margin of Pangea, rather than insulation. Whatever the cause, the formation and breakup of Pangea was clearly a time of major changes in the earth's geoid, changes that may also complicate calculations of sea-level fluctuations.

Permian Sea-Level Changes

A marine regression is associated with each of the major mass extinctions (Jablonski 1985; Hallam 1989a), although the nature of the connection remains uncertain. Fluctuations in sea level occur on both regional and global scales, and range from glacially induced changes over several tens of thousands of years to longer-term variations in the volume of mid-ocean ridges as the rate of sea-floor spreading changes. Changes in mantle convection may have significant impact on sea level as well (Gurnis 1992). Furthermore, short-term cycles are superimposed on long-term cycles, which are in turn superimposed on long-term aperiodic trends. Reconstructing changes in sea level can be difficult, but is relatively straightforward compared to identifying the source of the changes.

Three techniques are used to estimate the magnitude of variations in sea level: plotting changes in the deposition of marine sediments across continents; analyzing facies patterns to determine depth changes within sections; and sequence stratigraphy, which relies on the recognition of coastal onlap and offlap cycles in sequences (Hallam 1984a).

The standard description of the end-Permian regression includes a long-term drop in sea level beginning near the end of the Early Permian, with superimposed short-term transgressive-regressive cycles with periods ranging from 1 to 4 million years and averaging about 2.5 million years (figure 5.2). The regression continued into the Upper Permian, accelerating near the Permo-Triassic boundary where sea level reached perhaps the lowest point in the Phanerozoic. The drop is variously estimated at 210 m (Forney 1975) to 280 m (Holser and

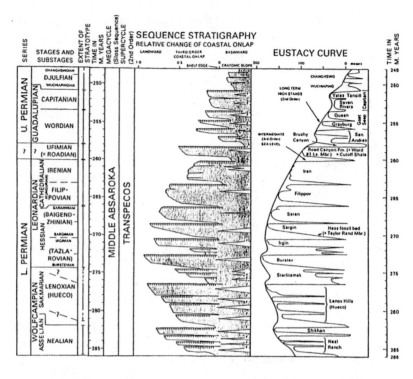

Figure 5.2. Permian sea-level curve inferred from seismic stratigraphy, lithostratigraphy, and biostratigraphy. Note the gradual decline from an Early Permian high, punctuated by shorter term (third-order) transgressive-regressive cycles. Adapted from Ross and Ross 1987.

Magaritz 1987). During the Early Triassic transgression, sea level rose rapidly by some 220 m. It should come as no surprise that with a regression of this magnitude preservation of marine sections across the Permo-Triassic boundary was limited. This considerably complicates efforts to understand the rate, timing, and pattern of the marine extinction.

One of the earliest attempts to quantify the extent of this regression involved plotting the distribution of epicontinental seas on paleogeographic reconstructions (Schopf 1974). Although the data were limited, Schopf demonstrated a sharp decrease in the percentage of the continents covered by shallow marine seas from 43 percent in the early Permian to about 13 percent at the boundary, followed by a rapid recovery to 34 percent in the Early Triassic.

Converting Schopf's calculated changes in the area of epicontinental seas to magnitudes of sea-level change requires determining the distribution of land area at particular elevations above sea level. The hypsometric curves resulting from such calculations can be used to determine the magnitude of change in sea level required to flood a given percentage of continental area, (Forney 1975). Assuming the hypsometric curve for the Permian was similar to today's, Forney estimated that sea level dropped about 210 m in the Late Permian, then rose 220 m in the Early Triassic. Modern hypsometric curves differ between continents, although they are broadly similar between the shelf edge and an elevation of 1 km. Africa is the exception, for it appears to be elevated by about 200 m relative to the other continents, probably as a result of hot-spot activity. If the variation in these continental curves is indicative of the possible range of variation in hypsometric curves during the Permian, they can be used to calculate the range of sea-level change that would produce 13 percent coverage of the continents at the Permo-Triassic boundary. Forney calculates the Late Permian drop was between 125 and 225 m. Forney also calculated that the volume of water displaced during a regression of 210 m was about 79.5×10^6 km^3. A cessation of sea-floor spreading for less than 10 million years, or a decrease in sea-floor spreading for a slightly longer period of time, could produce the 6 percent increase in the volume of the ocean basins necessary for a regression of this magnitude. Forney's assumption that the modern hypsometric curve can be applied to the Permian has been supported by some (Wyatt 1987) but rejected by others (Hallam 1984a; Cathles and Hallam 1991).

Absent from Forney's approach, however, is consideration of the relationship between mean elevation of a continent and the size of the continent. Since continental area and the average height of a continent appear to be related (Hay and Southam 1977; Wyatt 1984), the percentage flooding produced by a given increase in sea level depends upon the area of the block; the larger the continent, the greater the average height and hence the smaller the area flooded by a particular increase in mean sea level (Wyatt 1987). (Forney [1975] had neglected to adjust the estimates of sea-level change for isostatic rebound following the removal of the weight of the water from the continental shelves. This reduces the estimated vertical change in sea level by about 30 percent. Wyatt [1987] did correct for isostatic adjustment.) Wyatt modified Forney's approach by treating each continental block as an independent unit and demonstrated that the effect of sea-level change varies in a nonlinear manner, not linearly as assumed by Forney. Consider, for example, a sea-level increase of 100 m relative to modern sea level and a tectonic block with an average height of 500 m and an area of 11 x 10^{12} m^2. Using the modern hypsometric curve, organisms living between 0 and 50 m below sea level will experience a 60 percent reduction in available area, yet total area between 0 and 200 m will increase 125 percent! Since diversity is highest in shallow marine environments, diversity should decrease despite the transgression—the opposite of what most people assume. The biologic impact of a change in sea level clearly depends on the size of the block, the magnitude of the change, the position of mean sea level when it occurs, and the distribution of plants and animals.

Recall, however, Forney's observation that today Africa appeared to be elevated by about 200 m relative to the other continents and Anderson's discussion of the changes in the earth's geoid associated with the formation of Pangea. Together these considerations suggest that the appropriate hypsometric curve for Pangea may have been similarly elevated relative to the modern hypsometric curve.

Stratigraphic analysis detailed the history of short-term cycles through the Early Permian, and perhaps into the early Guadalupian (Ross and Ross 1985, 1987). Although no Milankovitch cycles (variations in orbital cycles) are known with a period similar to these Permian cycles, the characteristic asymmetric pattern of gradual transgression and abrupt regression supports claims that these cycles represent a continuation of the Carboniferous cyclothems caused by the Permo-Carboniferous glaciation.

Another way to evaluate the magnitude and scope of the end-Permian regression is to consider the depositional history of individual basins. Holser and Magaritz (1987) used Anderson's (1981) Permo-Triassic correlations across 68 sedimentary basins in such a study. They separated the formations into marine or nonmarine units based on fossil content; some units could not be reliably assigned to either category. Their results (figure 5.3) confirm earlier work, indicating a gradual regression beginning in the Asselian, then a rapid drop of sea level during the final stages of the Permian. Only 7 percent of all stratigraphic sections experienced marine deposition at the close of the Permian, leading Holser and Magaritz to estimate a maximum regression of 280 m. Holser and Magaritz assumed the present-day hypsometric curve also applied to the Permian. The Triassic transgression was among the most rapid of the Phanerozoic; the entire late Permian drop was recouped during the first two biostratigraphic

Permian-Triassic Sea Level Fluctuations

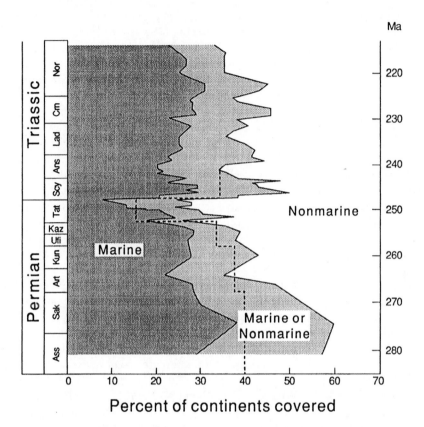

Figure 5.3. Permo-Triassic sea-level fluctuations based on analysis of 68 major sedimentary basins. The coverage of the continents dropped from about 40 percent in the Early Permian to 8–13 percent at the Permo-Triassic boundary. From Holser and Magaritz (1987: fig. 5) with permission of Gordon and Beach Publishers.

zones of the Triassic (the Lower Griesbachian Stage). If Holser and Magaritz's calculations are correct, this increase of as much as 280 m in sea level may have occurred in about 2 million years, or less, and ranks as one of the most rapid transgressions of the Phanerozoic.

Cathles and Hallam (1991) questioned Holser and Magaritz's results, pointing out that a regression of this magnitude should have left considerable evidence of erosion in boundary sections and produced abrupt jumps in stable isotope curves; they found no evidence for either. Cathles and Hallam are correct in that although there is evidence of nondeposition at the boundary, there is precious little evidence for erosion; however, the stable isotope curves suggest that the interval of nondeposition was considerable in some sections (Baud, Magaritz, and Holser 1989; see chapter 7). They also rejected reliance on modern hypsometric curves, albeit without addressing Forney's (1975) lengthy justification of these curves as a reasonable estimate of Permian hypsometry.

The new techniques of sequence stratigraphy provide a new way of thinking about sea-level change and new tools for studying the process, particularly through the identification of sequence and parasequence boundaries (see Hallam 1992 and Wilgus et al. 1988, respectively, for general and more technical introductions to sequence stratigraphy). Sequence stratigraphy relies on identification of unconformities to reveal the transgressions and regressions that bound depositional sequences. The earliest sequence stratigraphic analysis that included the Permo-Triassic boundary was based on seismic sections (Vail and Mitchum 1979). This study identified a lowstand at approximately the Lower/Upper Permian boundary, followed by a slight transgression in the Late Permian and a major rise in sea level in the Early Triassic. There is no independent support for this pattern of sea-level change. Vail and Mitchum were likely led astray by overreliance on seismic lines from North America, particularly from West Texas, where a regional regression began during the Guadalupian (see also Hallam 1984a; Ross and Ross 1987; Cathles and Hallam 1991). Wignall and Hallam (1992) have undertaken a preliminary sequence stratigraphic analysis of the Alpine boundary sections, but there is a compelling need for more rigorous sequence stratigraphic analysis of the boundary interval. Among other benefits, such work will allow a more precise definition of the magnitude of the latest Permian regression.

In summary, a variety of independent analyses produces results consistent with the experience of field geologists working in Permian rocks: a long-term gradual regression began during the Early Permian that was also a time of short-term transgressions and regressions. A slight recovery occurred during a Guadalupian transgression, but then sea level resumed falling, culminating in a very rapid regression at the close of the period. A rapid transgression then occurred during the earliest Triassic. Thus the mid-upper Permian regression includes two components: a gradual regression over at least 10 million years or more, and a very sharp sea-level drop of perhaps 100 m in 2 million years or less, followed by an equally rapid recovery. As we will see in the next section, explaining the long-term regression is relatively easy. However, the duration and magnitude of the boundary regression places it within the class of regressions that are among the most difficult to explain: too long to be driven by the formation of continental ice sheets, and too short to be clearly due to tectonic events.

Causes of Sea-Level Change

Global variations in mean sea level can result from changes in the volume of ocean water, the volume of ocean basins, in elevation of the continents, and other mantle processes (Southam and Hay 1981; Donovan and Jones 1979; Hallam 1984a; Worsley, Nance, and Moody 1984; Gurnis 1992). Taking each in turn, mechanisms for changes in the volume of ocean water include the formation of continental ice sheets and desiccation of ocean basins. The removal of water from the oceans to form continental ice sheets can produce changes of 150 m or more on time scales of thousands to tens of thousands of years, while the drying out of the Mediterranean caused only a 12 m global sea-level increase during the late Miocene (Donovan and Jones 1979). Related causes of change include variations in the volume of the hydrosphere, mean ocean temperature, and atmospheric moisture, but all appear to be insignificant and are safely ignored.

Changes in the volume of ocean basins have also received considerable attention, but such changes occur over far longer time scales than glacially driven sea-level fluctuations. Early in the development of plate tectonics geologists postulated a relationship between rates of sea-floor spreading and sea-level fluctuations: the depth of ocean crust is related to the age of the crust, for as it gets older it cools, becomes more dense, and sinks. Mid-ocean ridges with high spreading rates produce large areas of young crust, increasing the volume of the mid-ocean ridge system and decreasing the volume of the ocean basins. This causes a transgression by forcing water up onto the shelves and flooding the continents. Likewise, a decrease in sea-floor spreading rates should cause a regression (Sclater and Fancheteau 1970; Turcotte and Burke 1978; earlier literature is reviewed in Schopf 1974). For example, Hays and Pitman (1973) confirmed Hallam's (1971) hypothesis that a mid-Cretaceous increase in the rate of sea-floor spreading caused the mid-upper Cretaceous transgression.

In oceanic subduction complexes like the western Pacific, an increase in spreading rate may cause a decrease in sea level—the opposite of what most geologists expect. The actual change in sea level depends on the degree of cooling of the mantle induced by the subducted material (Hager 1980), as well as the plate velocity (Gurnis 1990). Initiation of subduction may also increase continental subsidence (Gurnis 1992) and cause a decline in sea level. Hager suggests that the change in volume of the ocean basins will equal that of the change in the volume of the mid-ocean ridges only if the mantle beneath the continents is cooled to the same extent as the mantle under the oceans.

Valentine and Moores (1970) first investigated the relationship between sea-floor spreading and the Permian mass extinction, suggesting that as sea-floor spreading declined following the formation of Pangea, the oceans deepened and the volume of the ocean basins increased. Later they concluded the relationship was more complex (Valentine and Moores 1972), with the extent of transgression and regression sensitive to changes in the topography of the mantle. For example, a continent adjacent to a rapidly spreading mid-ocean ridge might be depressed by the removal of material from the mantle underlying the continent needed to supply the ridge.

The volume of ocean ridges is controlled by variations in the length of spreading ridges, which will be at a maximum when there are a large number of plates and during periods of continental dispersion (Fischer 1984; Worsley, Nance, and Moody 1984; Worsley, Moody, and Nance 1985). The formation of Pangea should have been accompanied by a decrease in the length of spreading ridges and a regression (Valentine and Moores 1970, 1973). Despite the many complexities governing changes in sea level Valentine and Moores's intuition was basically correct. Over the past 180 million years there is a fairly tight connection between the subduction rate and global sea level about 30 million years later, as estimated from seismic stratigraphy (Engebretson et al. 1992). If the subduction rate did decline during the Early Permian following the initial formation of Pangea, this might explain some of the Late Permian regression. But the movement of the Asian microcontinents indicates that several spreading ridges persisted long after the initial formation of Pangea.

A novel mechanism to generate transgressive-regressive couplets over an interval of one million years or so has been advanced by Cathles and Hallam (1991) with particular reference to the end-Permian. They observe that the buildup of stress can induce changes in plate density that will propagate across the entire plate in less than 30,000 years, and they further suggest that stress at plate margins can result in sea-level changes of several meters and that the formation of rifts may cause a drop of up to 50 m. The stress associated with continental collisions may cause changes in plate elevation of up to 200 m. On large plates the resulting increase in the volume of the ocean basins could easily be sufficient to produce a substantial regression. For example, they show that a 50 m increase in water depth over a large oceanic plate affected by such stress translates into a global sea-level fall of about 10 m and a 200 m drop produces a global regression of about 38 m.

This mechanism, while in need of further study, shows great promise for explaining many of the rapid small-scale, nonglacial transgressive-regressive cycles found in the geologic record. The end-Permian regression is of the same temporal scale as the transgressive-regressive cycles studied by Cathles and Hallam. However, published estimates for the magnitude of the regression are about an order of magnitude greater than those produced by these stress-induced changes. Since no major continental collisions occurred at the Permo-Triassic boundary, Cathles and Hallam must not only identify a rapid rifting event, but must also convincingly demonstrate that the magnitude of the end-Permian regression was far less than previous estimates suggest.

The Obsky paleo-ocean, a failed rift in western Siberia, is one possible site of such a Late Permian rapid rifting event (Aplonov 1988; Demenitskaya and Aplonov 1988). This incipient ocean basin opened enough to produce a pattern of magnetic stripes. The long triangular-shaped rift underlies the present-day course of the Ob River and is about 200 km long with a maximum width of 300 km. It appears that the rift opened about the Permo-Triassic boundary, or during the Early Triassic. Aplonov (1988) cites a date of 220–230 million years ago for the opening of the rift, after the formation of the Siberian flood basalts, which occur in the same area (see below). There is, however, considerable uncertainty about the dates.

Deposition of sediments in the ocean may also cause changes in sea level, but these are at least partially offset by isostatic depression of the underlying crust and appear to be relatively minor (Donovan and Jones 1979).

Ultimately, many of the possibilities raised above are coupled to shifts in mantle convection patterns and resulting changes in the geoid. Similar shifts in the geoid may have occurred as the Pangean continental crust high over the geoid thickened by thermal expansion (see above; Anderson 1982) and Pangea stood high relative to the ocean basins. Overall, the anomalous regional differences in sea-level change during the Permian, showing simultaneous transgressions in some regions and regressions in others, may reflect the position of different areas with respect to regional geoidal highs and lows, rather than changes in the volume of the ocean basins or glacio-eustatic variations.

These changes in the geoid are related to the final mechanism leading to sea-level changes—variation in continental elevation. Relative sea-level declines may reflect an increase in continental thickness, thereby raising the mean elevation of the continent and changing the hypsometric curve, or perhaps through thermal doming caused by a mantle superplume. As discussed above, Africa appears to be elevated by about 200 m relative to the other continents. Africa appears to be nearly stationary relative to the mantle, and heat buildup beneath the continent has elevated it. The average elevation of a continent is related to its area (figure 5.4) and Pangea-type supercontinents should have had the highest mean elevation (Worsley, Nance, and Moody

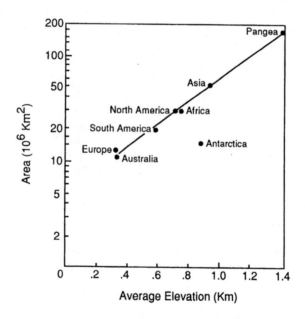

Figure 5.4. Relationship between continental size and elevation. From Worsley, Nance, and Moody (1984) after Hay and Southam (1977).

154

1984). Clearly this relationship may explain the long-term Permian regression but is insufficient to explain the rapid transgression and regression at the Permo-Triassic boundary.

Thermal uplift from the massive superplume that produced the Siberian traps may provide part of the answer. The volume of erupted material (1.5 x 10^6 km^2, according to Zolotukhin and Al'mukhamedov 1988; see below) is far larger than any other continental flood basalt and is surely only a fraction of the material contained within the mantle plume. Larson (1991b) estimated that the mid-Cretaceous superplume that produced the Ontong Java Plateau in the south-western Pacific elevated a region of 6,000 x 6,000 km^2. The increased thickness of continental crust would increase the amount of heat required to raise the same area, but the volume of the Siberian traps certainly suggests that this possibility should be further examined.

Strontium Isotopes

Another index of tectonic activity is the ratio of the two isotopes of strontium, although interpreting changes in these isotopes is not always straightforward (recall that isotopes have an identical number of protons, but differ in the number of neutrons). The ratio between the two isotopes of strontium, ^{87}Sr and ^{86}Sr, reflects the balance between two sources of strontium. Young basaltic rocks, like those on mid-ocean ridges, are enriched in ^{86}Sr, with an average value of 0.703. An increase in the rate of sea-floor spreading will enhance hydrothermal circulation through the fresh basalt and lower the oceanic strontium ratio. Conversely, continental granite has more ^{87}Sr. As continental crust forms through the fractional crystalization of mantle material, strontium is concentrated relative to both the mantle and oceanic crust, generating an average ^{87}Sr/^{86}Sr ratio of 0.710 (Holser, Magaritz, and Wright 1986), with older continental fragments even more enriched. As these rocks weather and erode the strontium isotopic ration of seawater increases. Thus the classical view of strontium ratios is that they reflect changes in the balance between continental erosion and runoff, which increases the ^{87}Sr/^{86}Sr ratio, and rates of sea-floor spreading and hydrothermal activity along mid-ocean ridges, which should lower the ^{87}Sr/^{86}Sr ratio.

Herein lies one of the many puzzles of the Permo-Triassic boundary. The strontium ratio drops from a relatively high level in the Early Permian to the lowest value of the entire Phanerozoic (0.70676; Kramm and Wedepohl 1991) near the Permo-Triassic boundary then climbs rapidly (Popp et al. 1986; Veizer 1989; Gruszczyński 1992). Geologic evidence indicates a marine regression during this interval, which is generally associated with a decline in the strontium ratio, but the strontium ratio went up, which suggests an influx of lighter strontium, perhaps from an increase in sea-floor spreading—which should cause a transgression. The reasons for this anomalous situation are not well understood, but Holser and Magaritz (1987) offered three possible solutions. First, if mid-ocean ridge activity were low, the strontium ratio may have been increased by generation of a major young, nonradiogenic, low-strontium continental input, perhaps like the Siberian flood basalts, followed by rapid erosion. The second possibility is a change in the rate of vertical mixing in the oceans, isolating the mid-ocean ridge input at depth, with riverine inputs added to the shallow reservoir, and rapidly changing the surface isotopic ratio. Oxygen isotopes provide no evidence for a stratified ocean in the latest Permian, however (see chapter 7). Finally, strontium may have been removed from the surface ocean reservoir, perhaps by the extensive mid-Permian evaporites, making the oceanic strontium isotopic ratio more sensitive to new additions. Of the three possibilities, the first appears to be the most likely, particularly in light of new dates placing the eruption of the Siberian traps close to the Permo-Triassic boundary (see below). This shift in strontium ratios coincides with marked changes in the isotopic ratios of carbon, oxygen, and sulfur, the significance of which will be considered in chapter 7.

The Kiaman Superchron

The Permo-Carboniferous Reversed quiet interval (PCR) (also known as the Kiaman Paleomagnetic superchron) came to a close near the end of the Permian and was replaced by the Illawarra mixed polarity interval. The PCR was a period of stable, reversed magnetic polarity (a compass needle would point to the south magnetic pole) that began during the Namurian Stage of the Late Carboniferous, or between the Namurian and the overlying Westphalian Stage about 323 million years ago and lasted until near the end of the Permian (Heller et al. 1988; Irving and Pullaiah 1976). Until recently, the upper boundary of the PCR was thought to lie very close to the Permo-Triassic boundary, although its placement varied at different localities. Recent work at Shangsi (Sichuan Province, China) indicates the entire Dalong Formation (of Longtanian-Changxingian age) belongs to the Illawarra superchron. Thus the PCR ended before the beginning of the Wujiapingian Stage in China or perhaps several million years before the Permo-Triassic boundary (Heller et al. 1988; Xu et al. 1989; Zakharov and Sokarev 1991; Haag and Heller 1991). Several magnetic reversals are recorded from the Upper Permian-Lower Triassic section at Shangsi: the upper Changxingian is normally polarized but another reversal occurs very near the boundary, followed by at least three more reversals in the Griesbachian (lowest Triassic) (Heller et al. 1988; Haag and Heller 1991; Steiner et al. 1989). The Illawara is followed by a mid-Triassic normal zone and then an Upper Triassic-Lower Jurassic mixed interval (Irving and Pullaiah 1976).

The PCR is one of two long intervals of stable magnetic polarity from the mid-Devonian to the present; the data are insufficient to determine if any occurred before the Devonian. The other is the Cretaceous Normal superchron, which ended immediately before the end-Cretaceous mass extinction. The congruence between the end of these two superchrons and mass extinctions has led to suggestions that the magnetic reversals may record changes in the internal structure of the earth, which in turn may be related to the mass extinction. These hypotheses will be considered further in chapter 8.

Late Permian Pyroclastic Volcanism

The 1980 Mt. St. Helens eruption, the 1991 eruption on Mt. Pinatubo in the Philippines, and the 1882 Krakatau eruption are all examples of pyroclastic eruptions. The global cooling following each event raises the possibility that larger pyroclastic events could have caused substantial climatic change and even mass extinctions. Most Permo-Triassic boundary sections in South China contain a series of altered volcanic ashes that mark nearby massive pyroclastic eruptions, eruptions orders of magnitude larger than anything experienced in recorded human history.

Pyroclastic eruptions are the most voluminous volcanic events known, with individual eruptions producing up to 4×10^3 km^3 of magma at rates of up to 10^6 m^3/second. The initial, plinian, phase in a pyroclastic eruption begins with the injection of a fragmented magma column to heights of 50 km or more. As the plinian column collapses widespread ash-fall deposits and pyroclastic flows form ash-flow sheets. Finally, an effusive phase may produce lava flows. In general, the larger the eruptive column, the larger the ash-flow sheets and the greater the amount of aerosols injected into the stratosphere. Fortunately for civilization, or unfortunately for curious volcanologists, no extremely large pyroclastic eruptions have occurred since the Toba eruption 75,000 years ago. The 1882 Krakatau eruption was one of the largest recorded pyroclastic eruptions, involving about 12 km³ of nonvesiculated magma erupted and ash reaching heights of 30 km (Simpkin and Fisk 1983).

The Permo-Triassic boundary clays in South China are altered volcanic ash 3–6 cm thick containing glass shards and bipyramidal quartz. Ash deposits cover an area of at least 3×10^6 km²; the estimated volume of erupted material is greater than 1,000 km³ (Zhou and Kyte 1988), which is equivalent to the entire yearly flow of the Yangtze River. Geochemical evidence clearly indicates the ashes were derived from a silica-rich pyroclastic eruption, probably from the subduction zones that rimmed Siberia during the late Paleozoic. The presence of bipyramidal quartz identifies an explosive volcanic event, not the more effusive volcanism generally associated with flood basalts. The end-Permian was also an interval of a different type of volcanism, however, the production of massive flood basalts.

FLOOD BASALTS AND MASSIVE VOLCANISM

FLOOD BASALTS are vast accumulations of basaltic flows produced by closely spaced eruptions. Typically the basalts cover the preexisting topography, leaving an extensive, flat-topped plateau. At the margins of the plateau, the different flows are frequently exposed as a series of giant steps. This terrace-like appearance lends flood basalts their other name, traps, from the Swedish word for stairs. Flood basalts are generally thought to be associated with the development of a mantle plume, or hot spot, and many flood basalts mark the beginning of continental rifting and breakup (Rampino and Strothers 1988).

The Siberian flood basalts of latest Permian age are one of the largest, and perhaps the largest, Phanerozoic flood basalt. The Siberian traps cover about 1.5×10^6 km² and range from 400 m to 3,000 m thick, including both intrusive and extrusive components (Zolotukhin and Al'mukhamedov 1988; Makarenko 1976). Since much of the flood basalt appears to extend toward the Ural Mountains beneath younger rocks, the total volume of the traps is 2–3 million cubic kilometers of material (Campbell et al. 1992). The Noril'sk section in the northwest section contains 45 flows in 11 sequences having a total composite thickness of 3,700 m (Renne and Basu 1991) and is the thickest part of the sequence. A smaller flood basalt province of Late Permian age in western China covers some 0.3 km^2 to depths reaching 2 km (Yin et al. 1989). Comparatively little is known about the Chinese flood basalt, but it may be contemporaneous with the Siberian traps and indicates the extent of heating beneath the supercontinent.

The age of the Siberian flood basalts and the interval during which they were erupted has been the subject of considerable controversy. Earlier studies suggested that the traps were erupted over a considerable interval from the Late Permian well into the Triassic (Sukhov and Golubkov 1965). Published radiometric dates range from about 290 to about 160 million years ago, but the lack of erosion between adjacent units within a vertical sequence suggests the entire sequence was erupted over a fairly short time span (Zolotukhin and Al'mukhamedov 1988; Baksi and Farrar 1991). Recent radiometric studies of the Siberian Traps give values clustering about 248 ± 4 million years (Campbell et al. 1992; Baksi and Farrar 1991; Renne and Basu 1991). (Initial differences in the age of the traps, as reported by Baksi and Farrar [1991] and Renne and Basu [1991] were caused by differences in analytical technique, as shown by Campbell et al. [1992].) Furthermore, the eruption of the Siberian Traps was very rapid. The entire sequence appears to have erupted over about 600,000 years (Campbell et al. 1992), more rapidly than previously suggested (Renne and Basu 1991). The eruption of the traps was essentially contemporaneous with the age of the Permo-Triassic boundary in South China (within the limits of analytical techniques). However, paleomagnetic evidence suggests the earliest flow, known as the Ivakinsky suite, erupted during a period of reversed polarity at the boundary itself. The remainder of the flows occurred during the Illawarra magnetic reversals of the earliest Triassic. The newly calculated eruption rates are about twice those of the Deccan Traps in India, the Karoo Province in South Africa, and the Columbia River Basalts. However, biostratigraphic evidence suggests that a portion of the upper unit of the flood basalt erupted over a longer period of time (Zolotukhin and Al'mukhavedov 1988; Sukhov et al. 1965).

As noted above, flood basalts are formed by either crustal rifting or the development of a mantle plume; in the latter case, a hot-spot track generally connects the flood basalt to an active hot spot. Morgan (1981) provisionally linked the Siberian traps with the Jan Mayen hot spot, presently north of Iceland (see also Renne and Basu 1991). This claim rests on a volcanic origin for the Lomonosov Ridge in the Arctic

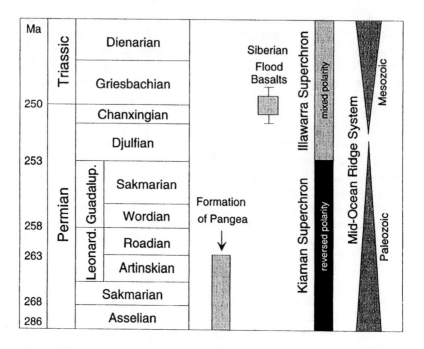

Figure 5.5. Summary diagram of major geologic and tectonic events near the Permo-Triassic boundary.

Ocean, but the ridge is rifted continental material, not volcanic in origin (Sweeney, Weber, and Blasco 1982; Jokat et al. 1992; Bill Holser kindly drew my attention to this discrepancy). The lack of a hot-spot track, and thus a mantle plume, associated with the Siberian Flood basalts is unusual, but not unprecedented. Too little is known of the western China flood basalts to speculate on their origin.

Figure 5.5 summarizes the geologic events during the Permian and Triassic. Clearly this was an interval of widespread geologic change and these changes had important effects on global and regional climate, our next topic.

VI

Icehouses, Greenhouses, and Saltboxes

A s THE Permian began, the Gondwanan continents were locked in the deepest glaciation of the Phanerozoic, but by mid-Permian times this gave way to warmer climatic conditions that lasted into the Cenozoic. This remarkable transformation from the icebox of the Early Permian to the greenhouse of the Late Permian was responsible for widespread changes in terrestrial plants and animals, including the replacement of the Paleophytic flora by the Mesophytic flora. These changes largely reflected the assembly and movement of Pangea, but debate continues over the importance of topography, changes in atmospheric CO_2, and other factors as supporting actors in the drama. Of course, because Pangea persisted from the Lower Permian into the Jurassic, its presence was not directly responsible for the extinction. Nonetheless, these long-term climatic changes set the stage for short-term perturbations that may have been involved in the extinction. Fingering climate change, or any other mechanism, as the forcing agent behind the mass extinction requires identifying a change occurring over approximately the same time scale as the extinction—no mean task, considering that we do not even know how long the extinction lasted!

There are two approaches to reconstructing Paleozoic climates: inference from the diversity patterns of plants and animals and the distribution of climatically controlled sediments such as evaporites, and theoretical modeling based largely on physical principles (Parrish 1982). Each approach has strengths and weaknesses, and in either case correlation problems and difficulties in temporal resolution necessitate an uncertainty of a few million years. We will begin by considering the direct evidence of Permian climatic changes before turning to climate models and the forcing factors that may have controlled some of these climatic changes.

PERMIAN CLIMATES: DIRECT EVIDENCE

The Permo-Carboniferous Glaciation

THE PERMO-CARBONIFEROUS glaciation began in the latest Devonian (Fammenian) as the southward drift of Gondwana brought it closer to the South Pole. The first pulse is recorded in glaciogenic sediments of northeastern Brazil and South Africa. Most models for the formation of polar ice caps and global cooling accord a primary role to continental distribution, although modulation of atmospheric CO_2 by increased weathering rates in low-latitude orogenies may also have been important. Continental ice may be several kilometers thick, while sea ice is only tens of meters thick. It follows from this that global cooling is most effective when large continental ice sheets can develop near both poles. The lack of such continental areas at high northern latitudes during most of the Permo-Carboniferous glaciation limited the magnitude of global cooling, and, coupled with the high temperature gradient between high latitudes and the equator, may explain the curious lack of any mass extinction during the Permo-Carboniferous glaciation. A second pulse of glaciation occurred in the Early Carboniferous (Visean) and was followed by the initiation of the main glaciation phase during the Namurian Stage (about 325 million years ago) (Crowell 1978, 1983, forthcoming; Veevers and Powell 1987). The glaciation ended in the early Guadalupian, for a total duration of about 70 million years.

Throughout this interval, as Gondwana moved across the South

Pole, large continental ice sheets expanded and contracted, reaching their maximum extent during the latest Carboniferous and earliest Permian (Asselian) (although see Crowley and Baum 1991 for another point of view). The volume of the sheets declined rapidly late in the Early Permian, with the last vestiges found in Australia and Siberia during the Guadalupian (Crowell 1978, 1983, forthcoming; but see Dickins 1985, who claims that this view considerably overstates the extent of the glaciation). Dropstones, apparently of glacial origin, are known as young as the Kazanian (Guadalupian). The magnitude of sea-level changes during the Permo-Carboniferous indicates that these ice sheets were of the same volume as those of the Pleistocene. Certainly they did not cover all of southern Gondwana, as some geologists have claimed, but fluctuated in volume as the supercontinent moved across the pole (Crowell 1978, 1983, forthcoming; Ross and Ross 1987). The distribution of glacial sediments on Early Permian paleogeographic reconstructions illustrates that ice sheets may have reached as far north as 40°, equivalent to the maximum extent of Pleistocene ice sheets (Crowell forthcoming).

The pattern of sea-level change during the Permian was discussed in chapter 5, but it is worth reiterating that short-term sea-level fluctuations during the Carboniferous and Early Permian are recorded as transgressive-regressive cycles ranging from 1.2 to 4 million years long and averaging about 2 million years (Ross and Ross 1985, 1987). These third-order sea-level changes are among the most enigmatic: too long to be generated by known perturbations in the earth's orbit (the Milankovitch cycles) and shorter than tectonic fluctuations. Curiously, these transgressive-regressive cycles may have persisted into the lower Guadalupian, and perhaps even into the upper Guadalupian (Ross and Ross 1987). If these cycles do record fluctuations in glacial activity, then ice sheets persisted far longer than previously suspected. There is no direct geologic evidence for such glaciations and Ross and Ross admit that the correlations on which this suggestion is based are at best highly problematic. Frakes and Francis (1988) have made the intriguing suggestion that high-latitude ice may have persisted through much of the Phanerozoic, but again, there is no direct evidence in favor of this argument.

An alternative, but highly speculative scenario is that increased weathering of silicates exposed by the Appalachian and other late Paleozoic low-latitude orogenies reduced CO_2 and triggered global cooling and glaciation. Maintenance of life on earth for at least the past 3.5 billion years requires that the atmospheric and oceanic carbon reservoirs must have been maintained within fairly narrow limits, so the rate of long-term mantle degassing of CO_2 must have been balanced by the removal of CO_2 from the atmosphere (Worsley and Kidder 1991), otherwise too much CO_2 would have tipped the earth into a runaway greenhouse like Venus; too little and temperatures would have plunged. Preventing global warming requires a negative feedback system in which weathering of silicates increases with increasing temperature, increasing the burial of CO_2 as carbonate, and weakening the greenhouse effect, while the opposite reaction occurs during global cooling. Worsley and Kidder (1991) recently proposed that periodic emergence and drowning of supercontinents has a primary role in long-term modulation of global climate. More specifically, they develop what they term a "2–10" rule for climate change: a decrease in emergent land area by a factor of 2 should increase atmospheric CO_2 levels enough to double the silicate weathering rate, producing a 10°C increase in temperature and balancing long-term CO_2 degassing and silicate weathering. They employ this relationship to construct a model in which land area and average latitude determine CO_2 values and global polar and tropical temperatures. Their model is highly simplified, but they find, for example, that mid-latitude ice caps are possible for emergent equatorial continents, contrary to received wisdom. Early results from this model are quite promising, but to simplify the model Worsley and Kidder have intentionally ignored a variety of important factors, including changes in the hypsometric curve, paleogeography, and weathering rates.

Nevertheless, there are several difficulties in applying this model to the Permian. Pangea was largely emergent during the Permo-Carboniferous and evenly distributed across the equator, yet no glaciation occurred. The increase in emergent land area during the Late Permian was associated with an increase in atmospheric CO_2 levels, the opposite of that postulated by Worsley and Kidder. Consideration of weathering and uplift rates suggests a solution to this dilemma.

Maureen Raymo and colleagues have examined the importance of weathering rates for global climate. They argue that the severe monsoons and steep slopes generated during uplift of mountains and plateaus enhances the rate of chemical weathering, thus decreasing atmospheric CO_2 (Raymo, Ruddiman, and Froelich 1988; Raymo 1991; Raymo and Ruddiman 1992). Uplift leads to mechanical breakdown of rocks and exposes greater surface area to chemical degradation. Carbon dioxide in the atmosphere combines with water to form carbonic acid, which in turn attacks silicate rocks. The bicarbonate ions released by the weathering process move down rivers to the oceans, where they are used to form shells and eventually deposited in sediment. The net effect of the process is to move carbon from the atmosphere into the sedimentary reservoir. Increased rainfall in mountainous regions also aids the weathering process. As noted previously, at some point global cooling slows the rate of chemical reactions, inhibiting further weathering and stabilizing CO_2. Since increases in the $^{87}Sr/^{86}Sr$ ratios roughly correspond to increases in the rate of continental erosion relative to sea-floor spreading, as described in chapter 6, strontium isotopes should serve as a proxy for global CO_2 levels (Raymo 1991): increased $^{87}Sr/^{86}Sr$ ratios reflect increased weathering rates and decreased CO_2 levels. The Phanerozoic strontium curve is consistent with this model during the Permo-Carboniferous glaciation, as well as the other major glacial episodes. The puzzling increase in $^{87}Sr/^{86}Sr$ ratios across the Permo-Triassic boundary would be interpreted as a decline in weathering rates, an increase in atmospheric CO_2, and global warming, a model for which there is considerable support.

Late Permian Climates

The abundant mid- to Late Permian climatic indicators, including coals, evaporites, reef carbonates, phosphorites, widespread terrestrial red beds, and climatically distinctive plant and animal fossils corroborate the establishment of a long-term warming trend following the glaciation (figures 6.1 and 6.2). The warming trend was intensified by the northward movement of Pangea into tropical latitudes and the expansion of the dry continental interiors as Pangea coalesced.

Permian red beds developed in association with dune, evaporite, and fluvial environments in low-latitude arid belts (Waugh 1973). Like evaporites, red beds are generally viewed as an indication of a hot, dry climate but the emphasis should be on the dry rather than the hot, since red beds can form at high latitudes in settings dominated by seasonal rainfall and drought, of which the Permian red beds of Australia are one such example. Upper Permian coals are restricted to high latitudes, unlike earlier intervals when they also developed along the equator (Parrish, Parrish, and Ziegler 1986) and are well developed in Australia, Antarctica, India, Siberia, and North China. In the marine realm the distribution of reefal carbonates can be one of the best climatic indicators, since modern reefs, dominated by scleractinian corals, occur only in tropical regions. However, Permian

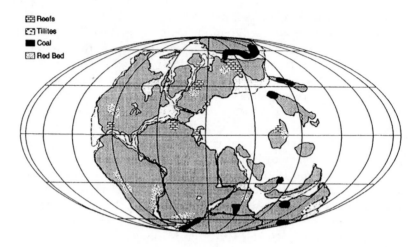

Figure 6.1.   Sedimentary climatic indicators for the Late Permian. This map and figures 6.2, 6.3, and 6.5 are the 250-million-years-ago reconstructions described in chapter 5. Data from a variety of sources.

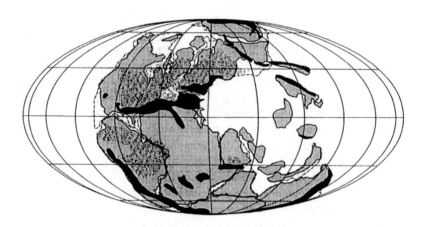

Figure 6.2.   Approximate distribution of mountains and highland areas (in black) and deserts (in stipple) during the Late Permian. Data from a variety of sources.

reefs were formed by a variety of different organisms, including tabulate and rugose corals, bryozoans, and sponges, and it is difficult to be sure that they were similarly restricted to the tropics. Still, Permian reefs do appear to have been restricted to low paleolatitudes, as are modern reefs.

Many large continental (intracratonic) basins developed during the formation of Pangea. Rapidly subsiding basins in dry climates with restricted circulation to the open ocean were sites of massive evaporite deposition. Some of these evaporites may have formed in only a few tens of thousands of years. Zharkov (1981) recognized 97 Permian evaporite basins (37 of salts and 48 of sulfates) with a total volume of 1.5×10^6 km^3. The largest Early Permian deposit fills the East European Basin from the Caspian to the Black Sea with some 500×10^3 km^3 of halite and 100×10^3 km^3 of anhydrite (Holser and Magaritz 1987). Other Late Permian evaporites include major deposits in the Zechstein Basin of northwestern Europe and the evaporites of West Texas. Together these basins define a broad band of equatorial aridity extending from about 35°N to about 50°S paleolatitude (figure 6.3; Zharkov 1981). Permian evaporites, of whatever age, are dwarfed by those of the Triassic, which account for 22 percent of all Phanerozoic evaporites (Gordon 1975).

One might surmise that removal of so much salt from the ocean substantially reduced salinity. Indeed, such changes have figured in

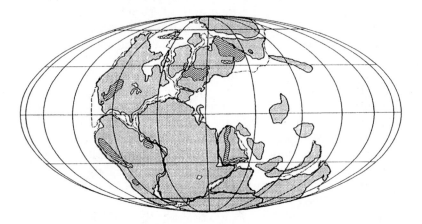

Figure 6.3. Approximate distribution of major zones of evaporite formation during the Permian. Data from Zharkov (1981).

a number of extinction scenarios (e.g., Fischer 1965; Lantzy, Dacey, and MacKenzie 1977; Stevens 1977). Yet two lines of evidence suggest that any salinity changes were minor. The first comes from sampling fluid inclusions preserved in Permian brines; the second from calculating the percentage of total modern salinity represented by modern evaporites and the effect on salinity if the maximum possible volume of Permian evaporites had been dissolved into the oceans.

Traditionally, geologists have relied on examination of mineral suites within marine evaporites to identify changes in the composition of seawater (see Holland 1984 for review) but the results are far from exact. A more accurate method has been developed by Heinrich Holland and his colleagues (Horita et al. 1991), involving analysis of small drops of the original fluids trapped within marine evaporites. These fluid inclusions are highly evaporated seawater, or brines, but their composition has been altered by diagenesis, dolomitization, and other chemical reactions. Consequently, the salinity of the original seawater cannot be determined directly. The concentration of bromine is relatively unaffected by these changes, however, and is directly related to original salinity. Fluid inclusions in Leonardian evaporites from Kansas and Tatarian evaporites from Texas indicate that the composition of Permian seawater was very similar to modern seawater (Horita et al. 1991). Evidently the extensive Permian evaporite deposits had little effect on marine salinity. As discussed in more detail in the following chapter, $\delta^{18}O$ varies with both salinity and temperature. The brine inclusions also confirm Holser's argument that an increase in the $\delta^{18}O$ signal in the upper portion of the Austrian GK-1 core indicates a temperature increase of 5–6°C (see chapter 7).

The effect of evaporite formation can also be estimated by calculating the impact of dissolving evaporites back into the oceans. For example, dissolving all the preserved halite (NaCl) in the rock record back into the oceans would only raise salinity by about 30 percent, which is the maximum (and highly unlikely) upper limit on Permian salinity (Claypool et al. 1980). The original volume of Permian evaporites may have been about 2.4 times the present volume of Permian evaporite basins, taking into account subsequent erosion, or perhaps 12 percent of the total amount of NaCl in modern oceans (Horita et al. 1991). If one fixes the amount of salts at the volume in Early Permian oceans, then deposit all the Permian evaporites, the maximum possible salinity drop during the Permian was about 12 percent. In fact, since older evaporites were almost certainly eroded during the

Permian, the actual drop in salinity must have been even less. Holser (personal communication cited in Horita et al. 1991) estimated on this basis Permian salinity changed less than 5 per mil, relative to about 35 per mil today. In short, the extensive Permian salts and sulfates withdrawn from the oceans were but a fragment of the total reservoir; evaporite deposition had little effect on salinity.

Evaporite basins may have had an important influence on marine circulation, however. When the temperature gradient between the poles and equator is low, warm saline water produced within evaporite basins and on shallow, low-latitude shelves, sinks rapidly into the deep ocean, inhibiting the circulation of cold polar waters and establishing a stable, stratified ocean with low-oxygen bottom waters (Holser 1984). Upwelling of cold, nutrient-rich waters is suppressed, in turn reducing primary productivity in surface waters. Evidence for warm deep-water brines and stratified oceans has not been produced for the Permian, but the conditions for their development did exist.

A Late Permian Glaciation?

Steven Stanley proposed that most major mass extinctions were caused by global cooling (Stanley 1984; see also Stanley 1988a, b). While studying Plio-Pleistocene bivalves of the Western Atlantic region, Stanley realized that glacial activity and global cooling had forced bivalves to migrate south along the Atlantic seaboard; species unable to move far enough south, because of the presence of Florida and the Gulf of Mexico, to stay within their optimal temperature zone had become extinct. The absence of impediments to southward migration along the Pacific coast of North America allowed bivalve faunas to simply move up and down the coast as temperatures waxed and waned. Stanley went on to examine the events surrounding other major mass extinctions and concluded that each was triggered by global cooling.

Most geologists approach hypotheses that wrap many complex historical events into a single neat explanation with some trepidation, for good reason, as history is sufficiently messy and the particularities of individual events sufficiently important that theories of everything almost always flounder. I was initially skeptical of Stanley's attempts to generalize this model to the Permian and other mass extinctions (Erwin 1989a), but I admit that by 1988 Stanley had marshaled enough convincing evidence for the Permo-Triassic that I later invoked global cooling as a contributory cause of the end-Permian mass extinction

(Erwin 1990b). Stanley's evidence includes dropstones and striated rocks, widely accepted indicators of past glacial activity, in Permian sediments from eastern Australia. Similar evidence in deep-water rocks (argillites) from the Kolyma tectonic block demonstrates a bipolar glaciation. The Kolyma block was an independent microplate during the Late Permian, but as Pangea moved north, the Kolyma block approached the North Pole and eventually collided with the Siberian platform during the late Mesozoic.

As Stanley (1988b) noted, the disappearance of reefs and carbonate sedimentation during the latest Permian and the depauperate, cosmopolitan fauna of the earliest Triassic provides supporting evidence for a latest Permian glaciation. For example, the calcisponge *Girtyo-coelia* and the encrusting alga *Tubiphytes* disappear in the Late Permian, but return along with reefs in the mid-Triassic. These taxa are sensitive to environmental disturbance, and their disappearance strengthens the case for environmental deterioration during this interval. The connection with global cooling is more tenuous, for the drop in carbonate deposition and the disappearance of reefs is also consistent with other extinction mechanisms, or it may simply reveal the effects of the extinction rather than its cause.

Finally, the continuation of the Carboniferous-Lower Permian transgressive-regressive cycle into the Guadalupian is equivocal, but supporting, evidence for glaciation. If the correlations proposed by Ross and Ross are valid, the global cycles could be interpreted as evidence of continuing glaciation.

Despite the apparent support for Stanley's hypothesis, it ultimately depends on the reliability of the latest Permian dates for the glacial sediments in Siberia and Australia. As it happens, convincing evidence, ignored by Stanley, has long been available that these sediments were deposited during the early Guadalupian, well before the end-Permian. They probably represent a final, bipolar phase of the Permo-Carboniferous glaciation triggered by the movement of Siberia and the Kolyma block into high northern latitudes. Once again, the importance of high-quality biostratigraphy and biostratigraphic correlation cannot be overstated.

Global climate models also suggest glaciation conditions suitable for the buildup of ice in both Siberia and eastern Australia during the Late Permian (Kutzbach and Gallimore 1989; Crowley, Mengel, and Short 1987; Crowley, Hyde, and Short 1989). These models also provide other interesting insights into global climate patterns during the Permian.

PERMIAN CLIMATES:
A THEORETICAL APPROACH

ATMOSPHERIC AND oceanic circulation are controlled by temperature contrasts between the equator and the poles, between the continents and the oceans, and by the rotation of the earth. As air near
the equator is warmed by the sun it rises and flows toward the poles,
transferring heat in the process. However, the Coriolis effect produced by the earth's rotation causes poleward-moving air masses to
be deflected eastward, generating clockwise atmospheric circulation
in the northern hemisphere, counterclockwise circulation in the
southern hemisphere. On an idealized earth with no continents,
atmospheric circulation would be divided into a series of cells parallel
with latitude, from the equator to 30°, 30° to 60°, and 60° to the poles.
Low pressure and high precipitation would develop at the equator,
and at 50°–60° latitude; high pressure and low precipitation at 30° and
near the poles. This combination of poleward heat transport and the
Coriolis effect produces a zonal circulation pattern (figure 6.4). The
introduction of continents disturbs these idealized circulation patterns, as do extensive mountain ranges and large upland areas such
as the Tibetan Plateau, that deflect normal circulation patterns.

Since continents have lower heat capacity than the ocean, they
heat up and cool off more rapidly. Continents warm during the sum-

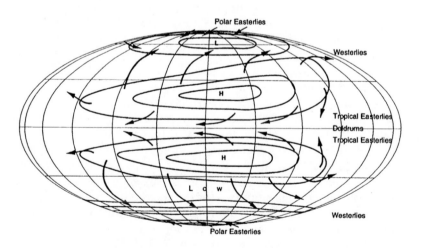

Figure 6.4. Idealized zonal circulation for an earth with no continents.

171

mer, heating the air mass above them and forming a regional low-pressure area as the overlying air masses rise; during the winter the continent cools, cooling the air mass, which sinks, forming high-pressure cells. These seasonal contrasts are described as continentality. The larger the continent the more isolated the continental interior from the ameliorating effects of the oceans and consequently the greater the seasonal contrast. When continentality and extensive uplands are combined, as they are in modern-day Asia, the result is a monsoonal climate with wet summers and dry winters. The core of Asia is isolated from the surrounding oceans by great distances, and further isolated by the presence of the Himalayan Mountains and the Tibetan Plateau along the southern margin and highlands to the east. These highlands act as a barrier to flow between the Pacific and Indian oceans and the interior of Asia. During the winter a very strong high-pressure cell develops with winds flowing out of the cell, while during the summer this is replaced by a low-pressure cell that draws warm, moist air masses off the Indian Ocean. The net effect is to produce a strong seasonal contrast and a regional disturbance in both atmospheric and oceanic circulation. These cross-equatorial monsoonal winds are the only winds that cross the equator. In doing so they destroy the independence of atmospheric circulation between the two hemispheres (Parrish 1982).

These general principles allow paleoclimatologists to reconstruct probable past atmospheric circulation patterns (figure 6.5). The formation of Pangea would have greatly increased continentality over what occurs today in Asia. The northern Guadalupian summer shows high-pressure systems off the western coasts of Laurasia and Gondwana and a pronounced low-pressure system developing slightly east of center in Gondwana. This Gondwanan high produced a dry winter, whereas the southwestern margin of Laurasia was subject to monsoonal conditions that may have been far worse than those occurring today in Asia (Parrish 1982; Parrish, Parrish, and Ziegler 1986; Parrish and Curtis 1982; see also Robinson 1973). The Appalachian Mountains formed a barrier between the interior of present-day North America and Europe and the Panthalassic Ocean, much as the Himalayas do today. Moreover, during the Mesozoic, unlike today, high plateaus in Gondwana permitted the development of monsoonal conditions in both hemispheres. Monsoons may explain such anomalous distributions of climatic indicators as evaporites in regions that today have high precipitation rates (Parrish, Parrish, and Ziegler 1986). Atmospheric circulation models also reveal regions where persistent

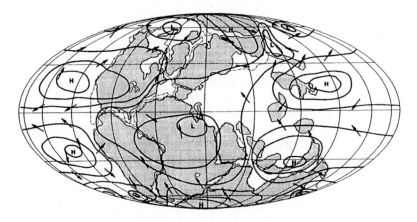

Figure 6.5. Reconstruction of Permian atmospheric circulation patterns, northern hemisphere summer. Based on Parrish, Parrish, and Ziegler (1986), Robinson (1973), and other sources.

winds parallel to coastal margins and associated with long-term high-pressure cells are likely to produce upwelling zones (Parrish 1982). Upwelling of nutrient-rich deep waters produces high primary productivity in surface waters; the resulting organic-rich sediments are an important source of petroleum.

Climate Models

The initial climatic reconstructions by Parrish, Robinson, and others can be extended by the development of supercomputer climate simulations. Such models are very sensitive to atmospheric composition, topography, and the distribution of clouds, but unfortunately none of these variables can be known with great accuracy. Nonetheless, we can make plausible estimates for these variables and test the resulting models against climatic indicators preserved in the rock record. Although many complicating factors are left out of these simulations, and the resolution is so broad that local and regional details cannot be resolved, simulations may reveal the interactions between different climatic variables.

Simulations allow paleoclimatologists to change the values and importance of variables built into the simulations, for example, the effect of topography or atmospheric composition. From these experiments, they can determine the relative contribution of each variable to global climate. Moreover, they can tune the models to match the

173

distribution of climatic indicators in the geologic record. This gives us a more general picture of global climate than can be achieved from climatic indicators alone.

The two most important classes of climate simulations are two-dimensional energy balance models (EBMs) and general circulation models (GCMs). EBMs determine global temperature patterns based on the balance between incoming solar radiation, diffusive heat transport within the atmosphere, and outgoing solar radiation as a function of the distribution of continents. These models are particularly useful in determining the effects of supercontinents on seasonality and temperature, particularly since EBMs can be run in a shorter period of time than the more complex three-dimensional GCMs. EBMs ignore oceanic circulation, topography, and other variables (Crowley and North 1991), but in general, results from EBMs correspond closely to modern climate patterns (Crowley, Mengel, and Short 1987).

Model runs of Upper Permian EBM simulations produce mean monthly temperatures in the interior of Pangea of more than 35°C, with daily highs of 50°C or more. Yet polar temperatures may have been low enough for the buildup of glacial deposits (figure 6.6; Crowley, Mengel, and Short 1987; Crowley, Hyde, and Short 1989; Crowley and North 1991; Crowley and Baum 1992). These results are broadly consistent with the distribution of climate-sensitive sediments, but the extremes exceed the temperature tolerance of most animals—in a region with widespread vertebrate deposits! Simulations also reveal extensive seasonality, with annual variations in temperature exceeding 30°C over much of Gondwana and over 50°C in the southern interior. Seasonal variations this broad are uncommon today, but do occur in regions of northern Canada and Siberia (Crowley, Hyde, and Short 1989). As noted earlier in this chapter, these results also suggest that high-latitude polar glaciation could have developed at this time. Indeed the northern migration of Pangea would make such glaciation more likely during the Late Permian. In the face of this, the lack of Late Permian continental glaciation demonstrates that some factor other than geography, such as an increase in atmospheric CO_2, must have acted to limit continental glaciation (see Crowley and Baum 1992).

General circulation models (GCMs) have also been used to investigate Pangean climates. This class of models divides the earth into a series of grids and, using partial differential equations covering the motions of the atmosphere and oceans, iteratively constructs a model of global climate. Unlike the emphasis on energy flux in EBMs, tem-

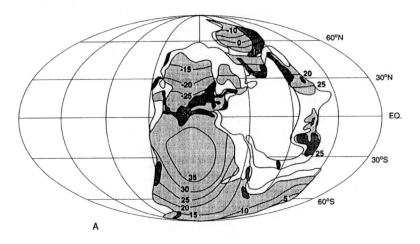

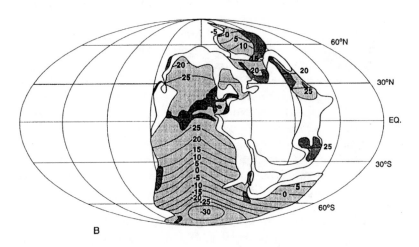

Figure 6.6. Results of EBM results showing possible temperature variations for the Late Permian: (a) northern hemisphere winter; (b) northern hemisphere summer. Open circles in Australia refer to possible continental ice. From Crowley, Hyde, and Short (1989).

perature, winds, and precipitation are all calculated by GCMs. GCM runs can be of two kinds: simulations, which attempt to simulate the climate of an interval of interest, and sensitivity experiments in which critical variables are altered to determine the response of the models (and, by inference, past climates) to change in these variables.

175

For example, paleoclimatologists interested in the Permian are particularly concerned with the effect of Pangea, so they change the land-sea distribution. Topography, atmospheric CO_2, and other variables may be examined in a similar fashion.

Kutzbach and Gallimore's (1989) studies of Pangean climates used a low-resolution atmospheric model coupled to a mixed-layer ocean model. The model includes parameters for sea ice and soil moisture but does not consider oceanic heat transport and circulation, CO_2 variation, changes in snow cover, etc. Because of uncertainties about several of the variables, rather than simulate Pangean climates sensitivity experiments were run with different values for topography, snow cover, and a combination of solar luminosity and CO_2 content. In general, the results of the sensitivity experiments are in broad agreement with the distribution of climatic indicators. They confirm the earlier conceptual models of Parrish and her colleagues (Parrish 1982; Parrish, Parrish, and Ziegler 1986) and the energy balance models of Crowley and his colleagues (Crowley and North 1991; Crowley, Hyde, and Short 1989). The GCMs produce strong monsoonal circulation, widespread aridity in continental interiors, equatorial aridity—unlike the present equatorial high-precipitation belt—and mid-latitude winter rain belts. High temperatures and extreme seasonal temperature variations were also generated, although Kutzbach and Gallimore caution that the model is likely to exaggerate the magnitude of these extremes by several degrees.

Kutzbach and Gallimore's cautionary note was prescient, for the recent discovery of a well-preserved forest from the Upper Permian of Antarctica demonstrates that despite the many advances made in computer simulations, the results of such studies may be far from reliable. Taylor, Taylor, and Cúneo (1992) discovered a fossil forest in the central Transantarctic Mountains, which includes 15 permineralized trunks in growth position. During the Late Permian this region lay at about 80° south latitude, and tree rings suggest the deposit is the remains of a rapidly growing young forest. The marked seasonality and the presence of deciduous trees suggests a climate more like that found today 10–15° closer to the equator. These results do not support paleoclimate models, including those discussed previously, with average winter temperatures of $-30°$ to $-40°C$ and summer temperatures of 0°C. The incongruence between the models and geologic evidence points to the fact that fossils and sediments remain more reliable indicators of climate history than the relatively primitive models.

In 1858 James Dana suggested that the late Cenozoic glaciation was linked to mountain building. Dana's hypothesis was championed by other scientists, including Charles Lyell, James LeConte, and Arthur Holmes, but finally fell from fashion with the discovery of multiple glacial expansions and contractions (see Molnar and England 1990, and references therein). These glacial fluctuations occurred more rapidly than conceivable rates of uplift and denudation. Recently Dana's idea has been resurrected with the proposal that late Cenozoic uplift of the Tibetan Plateau, Himalayan Mountains, Colorado Plateau, and the Western North America Cordillera had a significant impact on global and regional climate patterns.

The Tibetan Plateau is about one-third the size of the continental United States with a mean elevation of 4.5 km (about 15,000 feet), while the Colorado Plateau has a mean altitude of 1.5 to 2.5 km. Characteristic changes in plant fossils suggest rapid uplift of these regions during the past 40 million years, which Ruddiman and Kutzbach suggest (Ruddiman and Kutzbach 1989, 1991; Raymo and Ruddiman 1992; Ruddiman, Prell, and Raymo 1989; Kutzbach, Ruddiman, and Prell 1989; see also Manabe and Broccoli 1990; Crowley and Baum 1991) was a major cause of global cooling. To test this hypothesis, Ruddiman and Kutzbach modified a GCM to simulate the climatic conditions over the past 40 million years, varying topography to determine the influence of this parameter. Their results demonstrate the effect of plateau uplift on climate: as highland areas rise, the flow of low-altitude winds and the jet stream is diverted around the plateau. During summer the air over the plateau is heated rapidly and rises, much as it does over a continent, but the plateau intensifies the flow. As the air rises, the amount of water vapor it holds drops and seasonal rains begin. This produces the monsoonal rains discussed previously. The low-pressure cell over the plateau draws in air from the surrounding region. In winter, the system reverses, the cool air over the plateau sinks and descends off the plateau. The rising air over the plateau in summer leads to sinking of air masses over adjacent oceanic regions, connecting the regional climate over the plateau to global climate patterns. The results from GCMs predict regional climate changes associated with plateau uplift that compare well with the effects actually seen near the American Cordillera and in Southeast Asia.

Correlations of various kinds are frequent in geologic time, but correlations do not by themselves demonstrate a causal connection between two events: Ruddiman's hypothesis may suffer from a common defect: the chicken-and-the-egg problem. Molnar and England (1990) have asserted that the paleontologic, sedimentologic, and geomorphic changes used to infer uplift may simply reflect the change in climate, not the rate and timing of uplift. They do not deny that uplifts have occurred, or the interactions identified by Ruddiman and Kutzbach between mean elevation, climate, and weathering, but suggest that the feedback is more complex than generally acknowledged.

Nonetheless, there is substantial agreement between Ruddiman and Kutzbach's hypothesis and the events of the Carboniferous and Permian. As the continents collided to form Pangea, a number of mountain-building episodes occurred, including the Uralian Orogeny between the Russian and Siberian platform, and the Hercynian Orogeny that closed the proto-Atlantic ocean. The Hercynian Orogeny extended from the southern United States through Europe to the southern margin of Asia, as the northern Laurasian supercontinent collided with the southern Gondwanan supercontinent to form the Pangean superdupercontinent. These orogenic episodes produced extensive upland areas and mountain ranges.

Furthermore, the relationship between continental area and mean continental elevation discussed in chapter 5 suggests a mean elevation of Pangea of about 1.4 km. This would substantially increase the impact of orogenic activity on climate, exacerbating the effect that Ruddiman and Kutzbach documented (1989) for the late Cenozoic. This would also involve changes in weathering rates along the lines of the model proposed by Raymo (1991) and discussed earlier in this chapter.

There is sufficient agreement among these models to conclude that Pangea was subject to extensive monsoonal conditions accompanied by significant cross-equatorial winds. Summer temperatures in the continental interior were probably extreme, with high evaporation rates and a large arid region. In all likelihood, a similar arid region persisted in the equatorial zone. Middle- and high-latitude winters appear to have been quite cold on the continent. Despite the agreement between these models and many climatic indicators, so many variables are poorly constrained and the models so simplified that there is considerable room for further work.

Atmospheric composition is one of the important, but poorly constrained, variables for climate models. We are increasingly aware of the role played by greenhouse gases such as CO_2 in modifying our climate. By trapping heat inside the atmosphere rather than permitting it to radiate back to space, greenhouse gases act as an insulating blanket to warm the earth. As CO_2 levels increase, global warming begins; if the level decreases, cooling may ensue. Carbon dioxide appears to have been the primary greenhouse gas throughout earth history, and modulation of atmospheric concentration is a primary control on transitions between greenhouse and icehouse conditions. The changes in the global carbon budget discussed in the preceding chapter almost mandate a dramatic change in the volume of carbon dioxide during the Late Permian and Early Triassic. Estimating the magnitude of this change is difficult and no foolproof techniques have been developed, particularly for the Paleozoic. Modeling the relationship between atmospheric CO_2 levels and other geologic parameters is one means of approaching the problem.

Robert Berner has developed one of the most widely discussed models of Phanerozoic atmospheric CO_2 (Berner 1990, 1991). His results suggest that CO_2 levels were low during the Carboniferous-Early Permian, coincident with the Permo-Carboniferous glaciation, and increased rapidly during the late Permian (figure 6.7). Taken at face value, Berner's model strengthens the proposition that CO_2 levels play a primary role in controlling climate, but the uncertainties built into the model are considerable. For example, release of CO_2 by metamorphic degassing of carbonate rocks and organic material is his primary source of atmospheric CO_2. The values for CO_2 come from a model of the sea-floor spreading rate that Berner assumes can be derived from the record of sea-level change (Kasting 1992). Other variables in the model include the effects of temperature and land plants on weathering rates, and of temperature and geography on surface runoff. Obviously, values for these parameters are exceedingly difficult to determine for the Permian. More important, the time scale of these models is very coarse, suggesting changes occurring over tens of millions of years. Such changes might well explain the gradual warming during the Permian, but have little direct applicability to the extinction.

Carbon dioxide has garnered much of the attention as a greenhouse gas, but methane (CH_4) has played an equally significant role. Today methane is generated in rice paddies, landfills, and by hind-gut fermentation in ruminants and termites, and by organic decay in wetlands (Nisbet 1990). Recently several geologists have suggested that the introduction of methane into the atmosphere from deep-sea gas-hydrates may have been an important factor in modulating glacial-

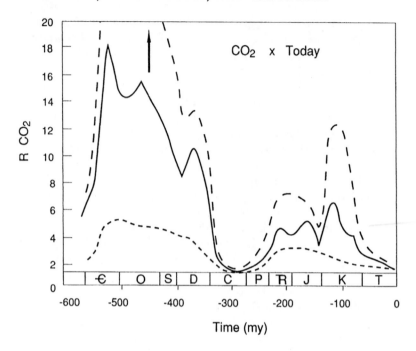

Figure 6.7. Berner's model of atmospheric CO_2 content through the Phanerozic. Note the lows during the Permo-Carboniferous glaciation and in the recent. From Berner (1991). Reprinted by permission of *American Journal of Science*.

interglacial transitions. The same mechanism may have played an important role in the end of the Permo-Carboniferous glaciation and in the initiation of the subsequent global warming.

The propensity of gas hydrates to form in, and clog, high-pressure gas pipelines has made these compounds the bane of the petroleum industry, but until recently little was known of their geologic distribution. These ice-like solids enclose methane, and sometimes other gases, within a water lattice and form readily in sediments where organic decay generates methane and where cold temperatures and pressure lock them into the sediment. They occur naturally under the permafrost and in sediments along the outer continental margin below 300–500 m, particularly in high latitudes and in regions where the permafrost persists offshore (Kvenvolden 1988). Deposits of gas hydrates are so ubiquitous that Keith Kvenvolden gave a minimum estimate of 11,000–16,000 gigatons (Gt) of carbon (1 gigaton = 1 billion metric tons) for total global hydrate reserves. By comparison, there are 600 Gt of carbon in the atmosphere (which may double in 50 years at the current rate), 5,000 Gt in all coal, oil, and gas reserves, or 833 Gt in all living things (Kvenvolden 1988). There is a lot of carbon in gas hydrates and even more carbon may have been locked up in gas hydrates during the Permian.

Before turning to the possible effect of gas hydrates on Permian climates, we should explore the effect of gas hydrates on the end of the recent ice age. Hydrates rapidly become unstable if temperature increases or pressure decreases, for instance during global warming or melting of the permafrost. Both marine transgressions and regressions can trigger releases, transgressions by melting the permafrost, for even Arctic Ocean water is above freezing, and regressions by releasing pressure on outer-shelf hydrates. Thus, the volume of the gas-hydrate reservoir is sensitive to both sea-level changes and to fluctuations in the size of the permafrost layer (Paull, Ussler, and Dillon 1991).

One model of the interaction between glaciation and hydrates, proposed by Charles Paull and William Ussler and William Dillon (1991; Paull and Ussler 1991), invokes the release of methane as sea level drops during the early stage of a glaciation. The initial release of methane into deep-sea sediments lubricates the hydrate zone, triggering slumps along the outer shelf, which in turn releases additional methane. Eventually the buildup of methane in the atmosphere causes global warming and limits the extent of glaciation and the volume of the ice sheets through a negative feedback loop (Nisbet 1990). For several reasons methane is a more efficient forcing agent for global warming than carbon dioxide. First, although methane has a lifetime of only 12–14 years before it is converted to CO_2, during that time it is a much more powerful greenhouse gas than CO_2. Second, the positive feedback mechanisms for the release of methane are greater than for CO_2 (Nisbet 1990). Deglaciations may be initially triggered by the release of massive amounts of methane, followed by a longer-term buildup of CO_2. As warming continues over thousands of years, old CO_2 will also degas from the ocean, further amplifying the greenhouse warming.

The volume of methane locked up during the Permo-Carboniferous glaciation is difficult to calculate, but almost certainly was far greater than the present global reservoir. Consider the extent of Permian epiric seas, and thus the areal extent of shelves within the gas-hydrate window of -300 to -500 m, the extensive area for permafrost development in southern Gondwana, and the magnitude of primary productivity during the Permian as evidenced by the volume of coal deposited. Together these factors must have generated considerable volumes of gas hydrates. Glacial fluctuations would have periodically released much of the marine hydrates and Nisbet's model of modulation of glacial cycles by methane release may well apply to the Permo-Carboniferous glaciation. However, as high sea levels until the mid-Permian confined most sea-level variations to the shelves, and even through regressions may have periodically exposed broad shelves, most of the gas-hydrate window would have been preserved. Thus by the Late Permian, sediments within the gas-hydrate window may have been saturated.

By the beginning of the Guadalupian, global warming and the northward drift of Gondwana would have freed most of the gas hydrates locked in polar permafrost, leaving the major reservoir along the shelves. As the Late Permian regression began, the reduction in the height of the water column removed the pressure keeping the gas hydrates in the sediment. The release of this methane would have further exacerbated the climatic effects of the increase in CO_2 and the decline in O_2 caused by the oxidation of organic matter on the exposed shelves. In time the methane would convert to CO_2 but would be replenished by the continuing marine regression. Glacio-eustatic regressions may be a self-limiting negative feedback loop if Nisbet (1990) is correct (see Lashof 1991 for a discussion of feedback loops, methane, and climate change). The global warming generated by methane release melted continental ice sheets, reversing the regression, and ending the buildup of methane and CO_2 in the atmosphere. Since the Late Permian regression appears to have been tectonically controlled, rather than a product of glacio-eustatic events, no such negative feedback loop would exist, and the methane release and resulting global warming would have continued until either the regression ended for other reasons, or the reservoir of gas hydrates was exhausted. If current estimates for the magnitude of the regression are accurate (and remember that considerable work is required to demonstrate this), then it appears likely that the gas hydrate reservoir may have been completely depleted. Other effects of methane-induced global warming have recently been discussed in conjunction with the early Eocene (Sloan et al. 1992). These links between methane and climate change in the Permian are frankly speculative. We have no direct evidence for the model described above, but it is consistent with our knowledge of the gas hydrates and their present distribution.

In summary, it is clear that we can draw limited conclusions about changes in atmospheric composition. As discussed in the following chapter, the exposure of the broad continental shelves during the regression — assuming estimates of regression magnitude are correct—and the resulting oxidation of organic material would have withdrawn oxygen from the atmosphere and increased the amount of carbon dioxide. Exposure and release of gas hydrates may have increased the volume of methane and subsequently carbon dioxide in the atmosphere.

Changes in greenhouse gases spring to mind as the most obvious source of climatic change, given the likely effect of our present uncontrolled experiment, but equally severe climatic changes may also be induced by changes in oceanic circulation.

Mikolajewicz, Santer, and Maier-Reimer (1990) describe the effects of a 50 percent increase in atmospheric CO_2 on the thermal expansion of the oceans. In GCM experiments they discovered that the formation of deep ocean waters in high latitude was no longer possible. Today deep ocean waters are an important control on heat transfer between the poles and equator. Moreover, CO_2 is sequestered in deep ocean waters, and thus a reduction in deep-water formation may further increase atmospheric CO_2, establishing a positive feedback loop. The time span over which these changes occur is fairly short and unlikely to have been particularly significant over the whole of the Permian, but the simulation gives some idea of the complex nature of the interactions between atmospheric composition and circulation.

The formation of Pangea appears to have been responsible, either directly or indirectly, for both the Permo-Carboniferous glaciation and the transition from the icehouse earth of the Early Permian to the greenhouse earth of the Late Permian (figure 6.8). The Permo-Carboniferous glaciation was initiated and sustained by the presence of large continental masses near the South Pole. Perhaps the magnitude of the glaciation was affected by changes in the volume of atmospheric carbon dioxide brought about by tectonically induced variations in weathering rates. The combination of the high mean elevation of Pangea, the formation of broad plateaus and extensive mountain chains as the supercontinent formed, and decreased atmospheric CO_2 brought on increased weathering rates that magnified the glaciation. Other processes may have played a role in this transition,

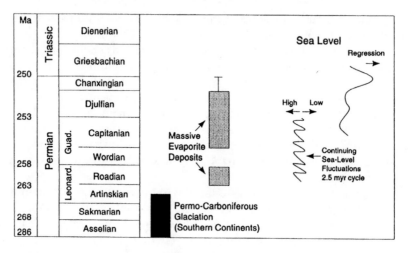

Figure 6.8. Summary climatic diagram for the Permian.

including the release of methane, or changes in the relative exposure of the continents, which changes the reflectivity (albedo) of the earth. The extent of vegetation may also have been important: when boreal forests are included in GCMs winter temperatures are warmer then if only snow and ice occur near the poles (Bonan, Pollard, and Thompson 1992).

The glaciation ended as Pangea moved into northern latitudes, but changes in atmospheric composition may also have been involved. Chief among these changes was the release of methane from the permafrost of Gondwana and from continental shelves as the regression began in the Late Permian. The buildup of these greenhouse gases exacerbated global warming in the mid-Permian as the bulk of Pangea reached low latitudes. By this time continental climates were hot, with large daily and seasonal swings in temperature and very severe seasonal storms. Evidence from the sedimentary and fossil record, and GCMs, all support continued global warming in the Late Permian; no evidence exists for the second glacial event near the Permo-Triassic boundary, as suggested by Stanley (1984, 1988a, b).

The interactions between geologic processes, climate change, and the history of life are complex, and commonly involve a variety of positive or negative feedback loops. Examples of these include the relationship between tectonic uplift, increased chemical weathering, reduction in atmospheric CO_2, a decline in global temperature that reduces the rate of chemical weathering and stabilizes CO_2. A more straightforward example is Nisbet's (1991) hypothesis linking glacial cycles to methane release: formation of continental ice sheets removes water from the oceans, lowering sea level and causing the release of gas hydrates. The greenhouse effect of the methane slows or reverses the ice formation, in turn slowing or reducing the regression. While the climatic effects and biologic consequences of these self-limiting feedback loops may be significant, they are inherently less severe than positive feedback loops that can trigger more far-reaching climatic change. Understanding the significance of positive feedback loops requires more information on chemical changes in the atmosphere and oceans, the topic of the next chapter.

VII

Geochemical Changes in
the Permian

THE CASTILE ANHYDRITE is a beautiful formation—hundreds of meters of thin, alternating light and dark bands of evaporites in the Delaware Basin of West Texas. The Castile was deposited near the end of the Permian, an interval marked by transition from the normal marine Bell Canyon Formation to the Castile. Within the lower few meters of the Castile, a marked shift occurs in the isotopic record of carbon. Such variations in carbon, oxygen, and sulfur isotopic ratios reflect changes in ocean chemistry and, through a complex network of feedback loops, in climate, biologic abundance, atmospheric composition, continental weathering, and oceanic circulation. Since each couplet of the Castile is interpreted as being laid down during a single year, geologists can count up the couplets and calculate the actual duration over which the Castile was deposited. By counting couplets and charting isotopes Mordechai Magaritz and his colleagues calculated that the carbon isotopes changed from $-2‰$

δ^{13}C to +6‰ δ^{13}C in less than 10,000 years (figure 7.1; Magaritz et al. 1983). This period of decreased oxidation of organic carbon continued until the Permo-Triassic boundary, where an equally marked negative drop occurred. Similar changes have been identified elsewhere, suggesting that the shifts may be global and can serve as valuable stratigraphic markers for correlating distant sections (Holser, Magaritz, and Clark 1986). Similar dramatic shifts in stable isotopes occur during each of the other major mass extinctions as well as the Cambrian-Precambrian boundary (Magaritz 1989).

Because the isotopic record may provide detailed information about climate, biotic abundance, and other factors closely related to the extinction, some understanding of what the numbers signify is important. We will return to the broader geologic implications of the record for climate in the following chapter, and for our understanding

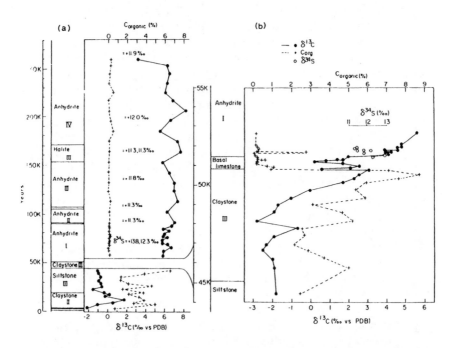

Figure 7.1. Onset of the increase in reduction of organic carbon in the transition from the Bell Canyon Formation to the Castile Anhydrite in the Delaware Basin, Texas: (a) the entire section, with scale in years on the left derived from counting evaporite varves; (b) δ^{13}C and δ^{16}O curves enlarged for the transition zone. From Magaritz et al. (1983).

of the Permo-Triassic boundary in chapter 9. (I freely admit to having a devil of a time remembering what all the jigs and jags of isotopic studies mean, so I have included as box 7.1 "The Compleat Idiot's Guide to Stable Isotopes." Readers need not feel insulted by the title; I wrote it for myself.)

BOX 7.1

The Compleat Idiot's Guide to Stable Isotopes

The element carbon is composed of three isotopes: ^{12}C, having six protons and six neutrons; ^{13}C, heavier with six protons and seven neutrons; and ^{14}C, with six protons and eight neutrons. Fortunately, ^{14}C is unstable and quickly decays to nitrogen, allowing archeologists to have a wonderful time dating charcoal and eliminating this isotope from further consideration here. Most carbon is ^{12}C, which is preferentially taken up during photosynthesis. In other words, the CO_2 used by plants to fix organic matter (giving off oxygen in the process) is enriched in ^{12}C relative to the atmospheric ratio. Burial of this organic carbon, whether on land, in shallow marine waters, or in the deep ocean, removes ^{12}C from the system and is recorded in the remaining carbonate as an increase in $\delta^{13}C$ (Kump 1992). Carbon isotopic ratios are subject to another confounding influence as well. In the oceans organic carbon (enriched in ^{12}C) sinks from the surface ocean to the deep ocean where it decomposes and is added to the dissolved carbon reservoir. The net effect of this carbon pump, as it is called, is to establish a gradient between the surface and deep ocean in $\delta^{13}C$. If primary productivity were reduced or eliminated at the surface (as may have occurred during the Cretaceous/Tertiary extinction), this gradient would decline until the $\delta^{13}C$ of surface and deep waters were the same.

Carbon is removed from the oceans in one of two forms, either as organic matter or as sedimentary carbonate particles such as calcite, aragonite, or dolomite (a magnesium carbonate). Fortunately, the isotopic ratios of organic matter remain unaffected by decay. As sedimentary carbonates are formed the ratio of ^{12}C to ^{13}C reflects that of the surrounding water. Thus carbonates record the original isotopic ratio.

Carbon is ultimately derived from the mantle, and if all mantle-derived carbon were removed as organic matter, or all as carbonate the ratio of ^{13}C to ^{12}C would be the same as the primitive ratio of the two isotopes as derived from the mantle. Since this is impossible, the changing ratios of ^{13}C to ^{12}C chronicle changes in the proportions of carbonate and organic carbon burial. As it happens, the primary source of variation in the isotopic ratio is variations in the burial rate of organic matter. Thus, if rocks from a particular interval are enriched in ^{13}C, this indicates that more organic carbon was buried. By analyzing changes in the ratio of the two isotopes through a section of rock, remembering that is the ratio in shallow water, changes in the relative rate of burial of organic carbon and carbonate may be estimated.

By convention, the ratio of the two elements is reported as $\delta^{13}C$, which is the difference, in parts per thousand, between the ratio of the two isotopes in a given sample and the ratio in a known standard. The standard is a Cretaceous belemnite shell from the Pee Dee Formation in South Carolina. The equation is thus:

$$\delta^{13}C = \frac{^{13}C/^{12}C_{sample} - {}^{13}C/^{12}C_{standard}}{13C/^{12}C_{standard}} \times 10^3 \ (\text{‰, PDB})$$

Thus a high $\delta^{13}C$ value indicates high rates of organic carbon burial and conversely a low $\delta^{13}C$ marks a shift in the budget, with high rates of carbonate deposition.

Several complicating factors must be taken into consideration in analyzing $\delta^{13}C$ and in interpreting published results. Most importantly, since diagenesis can alter $\delta^{13}C$ values the most reliable results are achieved from taking very small samples from fossil shells, bone, or teeth after these have been checked microscopically to make sure they have not been affected by diagenesis. Many older carbon isotope studies used whole-rock samples and may combine diagenetic effects with changes in ocean chemistry. Studies showing marked shifts in $\delta^{13}C$ across lithologic boundaries are particularly likely to be produced by diagenesis.

THE GLOBAL CARBON CYCLE

IMAGINE IF you will the annihilation of all living things, conversion of their carbon to carbon dioxide, and the addition of this carbon dioxide to the atmosphere. What effect would this have on $\delta^{13}C$? Very little, as it turns out. All the carbon released would be much less than all of that released by burning fossil fuels in the past 200 years (Berner and Lasaga 1989). The reason is that most of the earth's carbon is tied up in sedimentary rocks, not in the atmosphere or in the bodies of living plants and animals. Consequently, even a mass extinction as severe as the end-Permian is unlikely to cause changes in the ratios of carbon isotopes of the ocean or atmosphere. Although living organisms encompass only a small percentage of the earth's carbon, they act as the fulcrum that transfers carbon between the two major reservoirs—the carbon locked up in sedimentary carbonate rocks, such as limestones, and the buried organic carbon sequestered in coals, oil, gas hydrates, and the like. For example, a tremendous amount of organic matter is buried in a swamp, and as the productivity of the ecosystem increases, more carbon will be buried. Conversely, when the climate warms and the swamps dry out, less organic material will be buried and some of the previously buried material may be eroded. None of this would affect the isotopic ratios of carbon if it were not for the fact that photosynthesizing organisms are particularly finicky about which isotope of carbon they use: they are fond of ^{12}C, and, on average, during photosynthesis the ratio of $^{12}C/^{13}C$ increases by about 25–28‰ $\delta^{13}C$, the fractionation is less for marine plants than for terrestrial plants (see box 7.1 for discussion of this terminology). Thus as plants take up carbon in organic form and leave behind carbon enriched in ^{13}C, $\delta^{13}C$ also rises in the marine carbonate rocks in which it is usually measured.

These processes are part of the broader carbon cycle of both organic and inorganic processes. The major geologic processes within the carbon cycle (figure 7.2) include: 1) weathering of silicate and carbonate rocks using CO_2 and releasing dissolved HCO_3^- into the groundwater; 2) carbonate mineral deposition in the ocean utilizing HCO_3^- transferred by rivers; 3) burial of carbonates, followed by their destruction at depth during magmatism and metamorphism, and the return of carbon to the atmosphere-ocean system; 4) burial of organic matter in sediments, followed by later release during weathering; and 5) thermal breakdown of organic carbon at depth followed by later oxidation

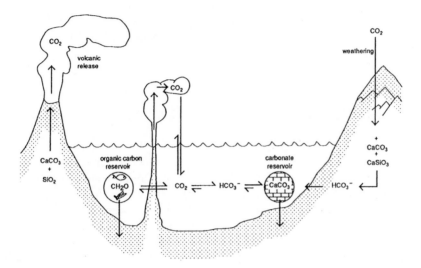

Figure 7.2. The global carbon cycle, showing the principal reservoirs.

of reduced carbon in the atmosphere (Berner 1990). On a geologic time scale, the lag between the production and burial of carbonates and their subsequent destruction during magmatism and metamorphism is critical. Although the weathering of carbonate minerals is important on a shorter time scale, it is generally disregarded when longer intervals are of interest. Changes in carbon ratios in the atmosphere and ocean reflect variations in the transfer of carbon between rocks and the ocean-atmosphere system over millions of years, but since the residence times of carbon within these two reservoirs are so short relative to geological processes, we can consider them as a single system. As discussed in box 7.1, it is the shifts in the relative proportions of the burial of organic carbon and carbonate that accounts for the fluctuations in $\delta^{13}C$ recorded through geologic sections. The isotopic shifts of $\delta^{13}C$ caused by increased storage of organic carbon will be globally synchronous within a few thousand years (Holser, Magaritz, and Wright 1986; Kump 1991).

The carbon and oxygen budgets of the earth are linked by the need for oxygen during the oxidation and weathering of carbon. Because the only source of atmospheric oxygen is photosynthesis, while the major sinks are respiration by animals and other nonautotrophic organisms, weathering of rocks, and the oxidation of gases released by volcanos and hydrothermal ridges, the amount of oxygen in the atmosphere is what is left over after respiration and geologic oxidation.

CHANGES IN CARBON AND OXYGEN ISOTOPES

THANKS TO the work of William Holser and his numerous collaborators, as well as other research groups, we now have a detailed understanding of the changes in carbon isotopes across the Permo-Triassic boundary at locations around the world. The $\delta^{13}C$ rose at the end of the Capitanian to $+3-+4‰$, apparently due to increased net organic storage of carbonduring the Late Permian. Close to the Permo-Triassic boundary $\delta^{13}C$ dropped to near 0, and in the best-preserved record of this interval there appear to be several positive and negative shifts near the boundary itself.

The increase in net reduction of organic carbon occurred over an interval of about 4,400 years in the Castile Anhydrite (Magaritz et al. 1983; figure 7.1), but so far this increase in $\delta^{13}C$ has only been identified in this area. Sampling through Western Tethys did not reveal this increase in net reduction of organic carbon and the GK-1 core drilled by Holser and colleagues in Austria (Holser et al. 1991) was not deep enough. Holser (1992) pointed out that the isotopic shift found in the Castile could be a short-lived excursion that they did not sample in Tethys.

A similar pattern of oxidation occurs in brachiopods from the Upper Permian (Artinskian-Changxingian) Kapp Starostin formation in West Spitsbergen (Gruszczyński et al. 1989; figure 7.3). Unlike the Castile, where the isotopic ratios may have been diagenetically altered—although steps were taken to eliminate this potential problem—the record from Spitsbergen comes from diagenetically unaltered brachiopod shells that accurately record the isotopic record of the shallow marine waters in which the animals lived. In Spitsbergen the $\delta^{13}C$ at the base of the section is about $+4‰$, consistent with other results for the mid-Permian, but then jumps quickly by $+3.5‰$. More interesting is the rapidity of the decline in $\delta^{13}C$ and $\delta^{16}O$ in the upper part of the section, with $\delta^{13}C$ falling by about 10‰ in perhaps a few million years. However, the Spitsbergen locality does not include the critical Permo-Triassic boundary interval, so it tells us little about the events close to the boundary. Similar drops in $\delta^{13}C$

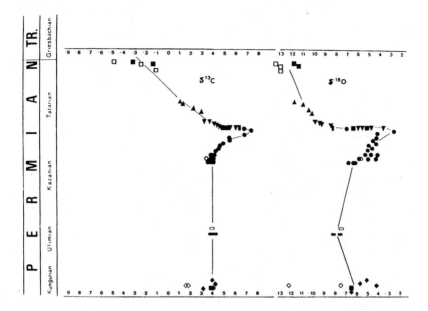

Figure 7.3. Stable isotope patterns for the Kapp Starostin Formation, Spitsbergen. Closed symbols are data from brachiopod shells; open symbols are other fossils and bulk rock samples. From Gruszczyński et al. (1989). Reprinted with permission of *Nature*.

have been recorded near the Permo-Triassic boundary in Iran (Holser and Magaritz 1987; figure 7.4) and throughout Tethys (Baud, Magaritz, and Holser 1989; Magaritz et al. 1988; figure 7.5), Greenland (Oberhansli et al. 1989) and South China (for locations of all sections see figure 7.6).

In South China, $\delta^{13}C$ increases abruptly at the base of the Changxingian formation, perhaps correlative with the increase seen in the Delaware Basin, in Spitsbergen, and in the Zechstein Basin of Western Europe (Chen et al. 1984). The upper portion of the South China sections shows a decline in $\delta^{13}C$, but the abruptness of the decline varies considerably between sections (Baud, Magaritz, and Holser 1989; Gruszczyński et al. 1990), reflecting variations in sedimentation rates and the extent of the hiatus caused by the regression. The shift appears more abrupt in localities where the regression was less complete.

Holser and his colleagues have undertaken an intensive investigation of a core through an apparently continuous Permo-Triassic

193

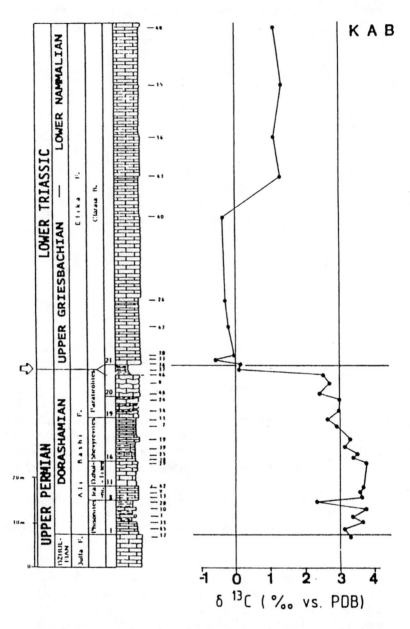

Figure 7.4. Carbon isotope data from the Kuh-e-Ali Bashi section in northwestern Iran. The rapid drop in the curve is produced by a condensed section in the upper part of the Dorashamian and a hiatus at the Permo-Triassic boundary, denoted by the arrow. From Baud, Magaritz, and Holser (1989). Drawn by A. Baud, samples deposited at Geological Museum, Lausanne, Switzerland.

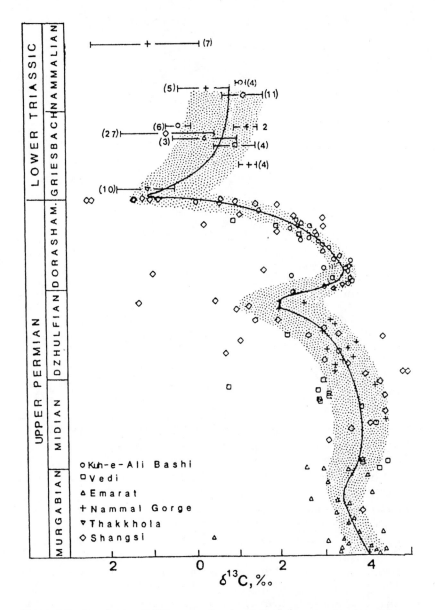

Figure 7.5. Synthesis of δ¹³C curves for six sections from western to eastern Tethys for the Late Permian and Early Triassic. The Permian curve is the authors' best estimate of the carbon isotopic curve for Tethys, with the shaded zone showing the degree of uncertainty around the estimate. From Baud, Magaritz, and Holser (1989), drawn by W. T. Holser.

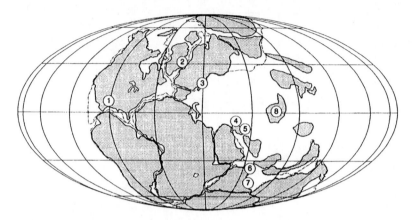

Figure 7.6.   Locality of Upper Permian-Lower Triassic sections with detailed $\delta^{13}C$ curves. (1) Delaware Basin, U.S.; (2) Jameson Land, Greenland; (3) Carnic Alps, Austria and Tesero, Italy; (4) Transcaucus localities: Kuh-e-Ali Bashi, Iran and Vedi and Sovetashen, Armenia; (5) Elburz Mountains, Emerat, Iran; (6) Nammal Gorge, Salt Range, Pakistan; (7) Thakkhola, Himalayas, Nepal; (8) South China (numerous sections). Based in part on figure 1 of Baud, Magaritz, and Holser (1989).

marine section in the Austrian Alps (Holser et al. 1989, 1991; figure 7.7). This core was taken from sections close to those discussed by Magaritz et al. (1988), but this Austrian GK-1 core is more complete and the section less condensed. The $\delta^{13}C$ record from the core confirms the pattern seen in the Delaware Basin: net reduction of organic carbon during the Late Permian followed by a gradually accelerating decline in the $\delta^{13}C$ curve. The GK-1 core also displays complex changes in trace element chemistry. For example, iridium exhibits two peaks in concentration above background in unit 3A that are associated with high concentrations of sulfur, iron, and other trace elements. These iridium concentrations are about one-tenth lower than those at the Cretaceous-Tertiary boundary and appear to be a result of biologic activity associated with the extreme regression. There is no evidence of an extraterrestrial source for this iridium. The nickel/iridium, cobalt/iridium and chromium/iridium ratios are all very different from either chondritic meteorites or the Cretaceous/Tertiary boundary clay. Rather, the trace element geochemistry and deposition of pyrite near the top of unit 2 suggests the development of anoxic conditions as the cause of the iridium concentration.

In the uppermost part of the Permian $\delta^{13}C$ declines continuously to a minimum near the boundary, with several additional minima in $\delta^{13}C$ in the Lower Griesbachian, followed by a return to a more normal $\delta^{13}C$ in the Upper Griesbachian, 1–3 million years after the extinction (Holser et al. 1991; Magaritz 1989). The multiple minima seen in the GK-1 core have not been observed in other sections (compare figure 7.7 with previous figures), perhaps because of the sections are more condensed elsewhere. The overall pattern is one of removal of isotopically light organic carbon, e.g., increased burial of organic

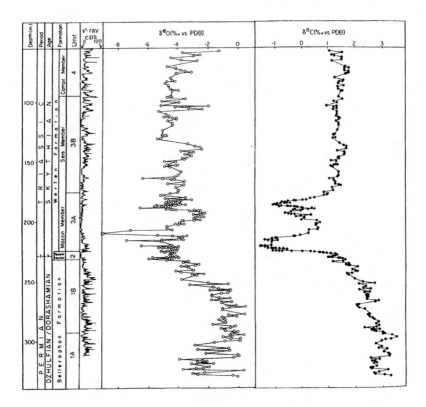

Figure 7.7. Carbon and oxygen stable isotope patterns, together with stratigraphy and geophysical log for the GK-1 core from the Carnic Alps, Austria. The Permo-Triassic boundary is in the lower portion of the Tesero Horizon (TH). From Holser et al. (1991). Copyright by Verlag der Geologischen Bundsanstalt, Wein, Austria.

carbon, followed by enrichment of light organic carbon either through reduced burial rates, increased input of light organic carbon from the land, marine deposits, or from volcanic carbon dioxide (Gruszczyński et al. 1989). The carbon isotopic record does not begin to return to normal values until the Late Griesbachian, 1–3 million years after the extinction (Magaritz 1989; Holser et al. 1991). This lag suggests that marine productivity and biomass were low for perhaps several million years into the Triassic, unlike the relatively rapid recoveries seen after the Precambrian-Cambrian and Cretaceous-Tertiary boundaries (Magaritz 1989).

There is some controversy regarding the interpretation of these isotopic excursions. Even though the general pattern is similar in all sections, the timing of the isotopic event relative to the Permo-Triassic extinction appears to vary, at least in part reflecting differences in the placement of the boundary (Oberhansli et al. 1989). There is another possible explanation for the isotopic excursions: the isotopic changes may be regional, reflecting changes in ocean chemistry associated with the formation of local evaporite basins and diagenesis. If this were so, the absolute values and the magnitude of the shifts should differ between basins, and we would not expect similar shifts in open marine settings. If the changes are basically indistinguishable except for differences in rate reflecting local variations in sedimentation rates, then the presumption must be that the isotopic changes reflect global changes. On balance, the isotopic signatures are so similar from sections ranging from restricted basins to open marine that the only reasonable conclusion is that the major shifts are globally synchronous events (see also discussion in Holser, Magaritz, and Clark 1986; Holser, Magaritz, and Wright 1986).

The $\delta^{18}O$ record of the GK-1 core is typical of other sections, dropping from about $-0.5‰$ in unit 1B of the Bellerophon Formation to about $-4.5‰ \pm 0.5$ at the boundary (figure 7.7). Oxygen isotope curves are influenced by temperature, formation of evaporites, and diagenesis. Holser et al. (1991) argue that diagenesis is not significant in this section. They suggest the cyclical changes in $\delta^{18}O$ in unit 1A at the base of the section reflect the formation of evaporites in the basin, but that the later, noncyclical changes were caused by increasing temperature. There is little direct support for this inference, but if they are correct, this drop signifies a temperature increase of about 6°C over an interval of less than 2 million years.

CHANGES IN SULFUR ISOTOPES

THE EXTENSIVE evaporites mentioned in the preceding chapter leave a signal in the sulfur isotopic record. Bacteria reduce sulfate to sulfide, forming pyrite in organic-rich muds. Weathering of the resulting shales oxidizes the sulfides, and the sulfates are then returned to the oceans. Sulfur precipitates from seawater during evaporation as anhydrite ($CaSO_4$) and gypsum ($CaSO_4 \cdot 2H_2O$) are formed. The heavy isotope of sulfur (^{34}S) precipitates only a little more readily than the lighter isotope (^{32}S), so evaporites will tend to act as a measure of seawater $\delta^{34}S$. The sulfur isotopic curve is derived from analyses of sulfate minerals in marine evaporites as deviations of $^{34}S/^{32}S$ from a meteoric standard, in parts per thousand, and reported as $\delta^{34}S$.

The sulfur isotopic record reached a nadir near the Permo-Triassic boundary, indeed the lowest point in the Phanerozoic, continuing a downward trend from a Lower Paleozoic high that began in the Devonian (Claypool et al. 1980; Kramm and Wedepohl 1991). The sulfur isotopic curve indicates a decreasing biological reduction of sulfate during the Upper Paleozoic. More interesting is the exceedingly rapid excursion of $\delta^{34}S$ from $+10.5‰$ to near $+28‰$ during the Spathian substage of the Scythian, perhaps in less than 500,000 years (Holser 1984). Kramm and Wedepohl (1991) described a fairly tight link between these variations in sulfur isotopes and the changes in strontium isotopes described in chapter 5. Recall that one of the possible explanations for the extreme shift in strontium isotopes was a decline in continental weathering and reduced input of weathering products into the ocean. A reduction in continental nutrients would affect the bacterial reduction of sulfates, and the correlation between the changes in strontium and sulfur isotopes suggests to Kramm and Wedepohl that there was a major reduction in continental inputs during the latest Permian. In addition, the production of carbonic acid may have occurred, which would have inhibited secretion of calcium carbonate.

IMPLICATIONS OF ISOTOPIC CHANGES

SIMILAR CHANGES in $\delta^{13}C$ occur at the onset of the metazoan radiation at the base of the Cambrian and at the Cretaceous-Tertiary boundary (Magaritz 1989), the mass extinction that wiped out the dinosaurs, made way for the radiation of mammals—and eventually the evolution of chocolate cake. Understanding the causes of the changes in $\delta^{13}C$ requires understanding the extent to which these variations reflect global variation rather than changes in fractionation, diagenesis, or other effects. The first step in this process is to determine $\Delta^{13}C$, which equals $\delta^{13}C_{carb} - \delta^{13}C_{org}$. If this value is relatively constant no significant changes in these other processes have occurred (Magaritz, Krishnamurthy, and Holser forthcoming). The GK-1 core provided the first opportunity to conduct such a study, which revealed little change in 'gD^{13}C, eliminating diagenetic effects or changes in fractionation as a cause of the changes in carbon isotopes (Magaritz, Krishnamurthy, and Holser forthcoming).

Given that these widely reported shifts in carbon isotopes are real, they could be caused either by the oxidation of living biomass or by a shift between the organic carbon and carbonate reservoirs, involving increased burial of organic carbon followed by its rapid oxidation. Following the methods outlined below, it is relatively straightforward to calculate the amount of biomass required to shift $\delta^{13}C$ by the amount observed. The amount required is about 6,500 to 8,400 gigatons of carbon (depending on whether the input is largely marine or terrestrial), far more than the 800 gigatons in living biomass. (After making these calculations I discovered that Gruszczyński et al. [1989]

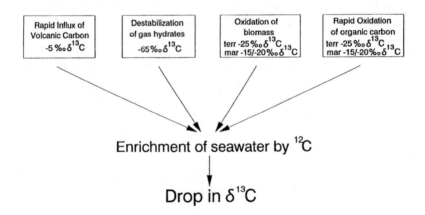

Figure 7.8. Possible sources of carbon to produce the shift in $\delta^{13}C$ at the close of the Permian.

had followed a similar procedure and arrived at the same conclusion.) Four processes could, in principle, be responsible for these events. All involve an extensive shift between the global organic and inorganic carbon reservoirs, either through a rapid influx of juvenile carbon in the form of CO_2 from volcanic gases, with a $\delta^{13}C$ of $-5‰$, the oxidation of a considerable volume of buried organic material, with $\delta^{13}C$ of -15 to $-20‰$, or release of gas hydrates as discussed in the previous chapter, with a $\delta^{13}C$ of $-65‰$ (figure 7.8).

As it happens, the volume of carbon required to produce a change in $\delta^{13}C$ from $+3°/oo$ to $-1°/oo$ is readily calculated using equations from Spitzy and Degens (1985) (table 7.1). These results reveal that the lower the $\delta^{13}C$ value of the carbon source the greater the volume of carbon required to produce the required shift. These calculations assume the changes occur relatively rapidly and ignore effects that will tend to reduce the effect of the carbon released. For example, addition of CO_2 to the oceans may raise the carbonate compensation depth (the depth at which carbonate skeletons dissolve) or to dissolve sedimentary carbonates. If the latter occurs, the heavier marine carbonates will reduce the volcanic signal by about half (Spitzy and Degens 1985). Other feedback systems will doubtless occur as well. Nonetheless, these calculation do provide a feel for the *relative* amount of different inputs required to shift oceanic $\delta^{13}C$.

Which of these are viable alternatives? To know that we need to understand the volume of different reservoirs within the carbon cycle, and the flux rates of such processes as volcanic release of CO_2. The volume of these reservoirs is shown in table 7.2. Caldeira and Rampino (1990a, b) estimated the total CO_2 release from the Deccan traps at about 1,000 gigatons. The volume of the Siberian traps was far larger than the Deccan traps and it is difficult to estimate the volume

TABLE 7.1 Volume of carbon from different sources required to produce a shift in $\delta^{13}C$ from $+3‰$ to $-1‰$ during the latest Permian. Volumes were calculated using the equations provided by Spitzy and Degens (1985) and for the amount of dissolved inorganic carbon in the Permian ocean—set at 40,000 gigatons, the value for the present ocean. Volumes in gigatons of carbon.

Volcanic gases	$-5 ‰ \, \delta^{13}C$	40,000 Gt
Organic carbon	-20	8400
Organic carbon	-25	6500
Gas hydrates	-65	2500

of CO_2 that may have been released. Nonetheless, it is clear that given the enormous volumes of CO_2 required to shift $\delta^{13}C$ to the extent required, the Siberian traps could not have contributed to more than a fraction of the shift (see also Gruszczyński et al. 1989). The second source for the isotopic shift is an increase in the delivery of bicarbonate ions from an increase in continental weathering. Again, the volume required relative to the expected flux is insufficient. The final two options are both viable, and during an extensive regression both are likely. The volume of the buried organic carbon reservoir today is probably less than would have been deposited in the the vast epicontiental seas of the Permian. Erosion of several thousand gigatons of organic carbon would seem quite feasible given the magnitude of the regression. Release of gas hydrates requires an even smaller volume of carbon released, in fact only a quarter of the (conservative) estimate of present-day reservoirs. Both oxidation of buried marine organics and gas hydrates is likely to accompany the marine regression.

These estimates assume the change in $\delta^{13}C$ reflected a whole-ocean shift. The other alternative is a decline in productivity in shallow marine waters as a consequence of the extinction and a decline in the $\delta^{13}C$ gradient between surface and deep waters. This has been described as a "Strangelove ocean" (Hsu et al. 1985) and may have occurred in the latest Precambrian and at the Cretaceous/Tertiary boundary. Any event that eliminates primary producers in surface waters must reduce the surface-deep ocean gradient caused by isotopic fractionation in surface waters.

TABLE 7.2 Volume of various carbon reservoirs within the global carbon cycle. Numbers are in gigatons of carbon.

Carbonate sediment reservoir	6,000,000
Organic carbon reservoir	1,300,000
Gas hydrates	10,000
Dissolved organics in oceans	1000
Oil, gas and coal	5000
Humus, peat, detritus	2000
Land biota	800
Marine biota	3
Atmosphere	700

From Kvenvolden (1988), Arthur (1982) and other sources.

The global ocean model (Holser et al. 1989, 1991), invokes the exposure of the continental shelves by the regression as a trigger for increased erosion and oxidation of buried organic matter. The drop in $\delta^{13}C$ roughly parallels the regression; recall, nevertheless, that a correlation suggests, but does not demonstrate, a causal connection. Oxidation of the organic matter reduced oxygen levels and increased the CO_2 content of the atmosphere, bringing about increased temperature and oceanic anoxia. Holser's argument (Holser et al. 1991) is strengthened by the discovery that continental shelves play a larger role in cycling of carbon than previously realized (Walsh 1991).

An alternative to Holser et al.'s model was proposed by Robert Berner, who suggested that warming and drying of the earth during the Permian would have reduced the rate of burial of organic matter, causing a decline in $\delta^{13}C$ and a reduction in atmospheric oxygen (Berner 1989b). Basically this amounts to the same thing as Holser's model but emphasizes climate change rather than the effects of the regression. Particularly significant in Berner's view is the disappearance of swamps and other moist lowland environments following the end of Permo-Carboniferous glaciation. These environments serve as the primary burial ground for organic matter and their loss via increasing continental aridity should have had a significant effect on carbon isotopes. I see no particular distinction, however, between Berner and Holser's approach, since both involve global shifts between the two carbon reservoirs, and I regard them as part of the same process (see Holser et al. 1991, fig. 13, for comparison). A related model has been advanced by another research group (Małkowski et al. 1989; Hoffman, Gruszczyński, and Małkowski 1990) and will be considered in greater detail in chapter 9.

How do we choose between the whole-ocean model and the shallow-water model? In post-Jurassic seas these alternatives are relatively easy to distinguish because of the presence of both pelagic (floating) and deep-water benthic foraminifera (and similar organisms with calcareous shells). Pelagic and benthic forams can be readily separated and isotopic analysis of both fractions reveals whether the shallow-water pelagic forms record greater fractionation of stable isotopes in surface waters than the benthic forms record in deeper waters. If biological productivity in shallow waters had largely disappeared the gradient in $\delta^{13}C$ between shallow and deep water would disappear as well. The difficulty in applying this approach to the Permo-Triassic record is that there are no obvious pelagic calcium secretors, and thus no apparent way to examine both the isotopic fractionation in both shallow and deep-marine organisms within a single section.

Magaritz and colleagues (forthcoming) recently suggested a way out of this dilemma. The organic material in sediments is another record of carbon isotopes as mentioned earlier (p. 200). This records the isotopic ratio of the organic reservoir much as the record from a limestone records the ratio of the inorganic carbonate reservoir. Because the atmosphere responds quickly to global ocean changes in carbonate, the resulting change in the volume of atmospheric CO_2 should change the extent of fractionation of carbon during photosynthesis. For example, a significant increase in CO_2 will enhance the enrichment of primary producers in C^{12} relative to C^{13} and this change in the extent of fractionation can be detected by examining the record of $\delta^{13}C$ from organic matter. In fact, analysis of C_{org} from the GK-1 core reveals no change in fractionation. These data are consistent with a drop in surface productivity across the Permo-Triassic boundary but not with claims for a global shift between the organic and inorganic carbonate reservoirs. The results would also appear to reject claims that the percentage of atmospheric oxygen dropped so low that this was a major factor in the extinction. It is less clear whether the lack of evidence for a change in fractionation also precludes an increase in atmospheric CO_2 sufficient to cause a greenhouse effect. The amount of CO_2 necessary for a greenhouse effect may be well below that which causes changes in C_{org} (Magaritz, Krishnamurthy, and Holser forthcoming).

CERIUM FRACTIONATION AS AN INDEX OF ANOXIA

MODERN OCEANS are generally well oxygenated at all depths, but there is considerable evidence that this has not always been the case. In the past, there have been considerable intervals during which substantial parts of the water column have been either anoxic (lacking oxygen) or dysaerobic (with anomalously low oxygen concentrations), as well as having more temporally restricted oceanic anoxia events (OAEs). These changes in marine redox conditions generate distinctive changes in the element cerium relative to other rare earth elements (REE) that provide a useful index for the development of anoxia. This is not the place to talk about the biotic effects of these episodes, which will be discussed in chapter 10, but because a major cerium anomaly occurs near the Permo-Triassic boundary it will be considered here.

The REE are arrayed down on the lower portion of the periodic table of the elements. Normally all of the REE except for cerium exhibit the same chemical behavior. Cerium is normally depleted relative to lanthanum and neodymium in oxidizing seawater and coprecipitates with metallic oxides (Wright, Schrader, and Holser 1987; Wright 1989). As the partial pressure of oxygen in the system drops, this cerium anomaly disappears and the relative abundance of cerium approaches that of the other REE. In the conventional notation for cerium abundances, the normal depletion of cerium produces a negative Ce_{anom} value; zero or positive values of Ce_{anom} occur under anoxic to dysaerobic conditions. (For those interested in all the nitty gritty details, the cerium anomaly is calculated as $Ce_{anom} = Log[3Ce_n/(2La_n + Nd_n)]$ where n is the shale-normalized value of an element [see Wright, Schrader, and Holser 1987 for discussion].) Although the values of Ce_{anom} vary considerably in the Atlantic and Pacific (Wright 1989), they are all negative. Exceptions are in the Gulf of California and the Black Sea, indicating that the buildup of cerium under anoxic conditions in deep water is reflected in shallower, oxic waters.

Studies by Wright (Wright, Schrader, and Holser 1987; Wright 1989) demonstrate that biogenic apatite (calcium phosphate) retains the signature of the redox conditions under which the skeletons formed. By analyzing variations in cerium values in conodonts and other phosphatic fossils from Cambrian through Jurassic carbonates, Wright produced a curve of the variations in Ce_{anom} (figure 7.9). This curve indicates that anoxic conditions predominated in lower Paleozoic oceans, but were replaced by increasingly oxic conditions from the Devonian into the Permian. The sharp change in Ce_{anom} across the Permo-Triassic boundary suggests a temporary return to anoxic or dysaerobic conditions. Normally oxidizing conditions returned later in the Triassic and continued into the Jurassic. One does not want too read to much into this curve. Examination of the original data (Wright, Schrader, and Holser 1987) indicates that the latest Permian samples available to Wright with normal (i.e., oxidizing) Ce_{anom} values were from the Guadalupian of Texas, whereas the anomaly shows up in Scythian conodonts from Utah and Nevada. Cerium values had returned to normal by the Norian, some 20 million years

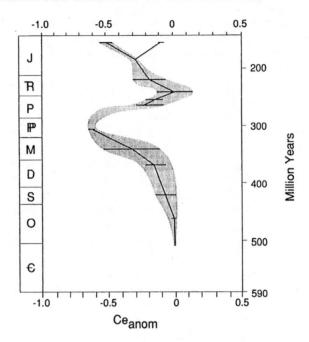

Figure 7.9. Cambrian to Jurassic cerium isotope anomalies. Horizontal bars are error estimates, and the shaded interval indicates the most probable range of the actual values of the cerium anomaly. From Wright (1989).

later. The most one can say from these results is that a fairly severe episode of marine anoxia appears to have affected the Scythian samples. The data do not reveal how close to the boundary this change occurred, nor do they exclude the possibility that only sections in western North American record the change. The possibility that this oceanic anoxic event triggered the end-Permian mass extinction has been advanced (Hallam 1991; Wignall and Hallam 1992) and will be evaluated in chapter 9.

VIII

Cycles of Change

ONE OF the enduring controversies in geology is between those who find in the history of the earth gradual and progressive geologic change and those who see the endless repetition of cycles, great and small. Lyell was an unabashed partisan for the latter point of view and converted many geologists to his side through the many editions of his *Principles of Geology*. More recently, catastrophists, largely banished from empirical science in the last century, have returned to respectability with the recognition that extraterrestrial impacts and their catastrophic effects have been a regular feature of earth history. Today, geologists recognize earth history as a complex web of unidirectional evolutionary changes, hierarchies of cyclical patterns, and random fluctuations, all periodically interrupted by catastrophic events of varying magnitude, some of extraterrestrial origin. The relative importance of the various elements of this web are a source of enduring controversy.

The formation of Pangea divides two very different eras in geologic history, much as the Paleozoic-Mesozoic boundary separates two distinct intervals in biologic history. This geologic transformation is reflected in changes in tectonic style from accretion and extensive orogeny to dispersal, in the geoid and the distribution of hot spots, in the end of a lengthy period of stable reversed magnetic polarity, in the onset of a period of fluctuating polarity, and in other tectonic processes. These events in turn altered atmospheric and oceanic composition and induced the climatic transformation from a cool late Paleozoic globe to a warmer one in the Mesozoic.

In 1924 Hans Stille returned to Lyell's cyclical view of the earth proposing a series of tectonic cycles driven by the intrinsic dynamism of the core and mantle. The episodic nature of orogenic events supports Stille's view, with the the formation of Pangea marking the transition from one cycle to the next (see discussions in Stille 1924; Umbgrove 1947; Holmes 1951; Sloss 1963, 1992; Johnson 1971; MacKenzie and Pigott 1981; Veevers 1989). In a particularly significant paper, Anderson (1982) suggested the insulating effects of supercontinents so retarded heat flow as to rearrange mantle convection patterns. Anderson's arguments have been central to many recent discussions invoking the changing dynamics of the core and mantle in controlling these periodicities (Fischer 1981, 1984; Worsley, Nance, and Moody 1984, 1986; Worsley, Moody, and Nance, 1985; Worsley and Nance 1989; Nance, Worsley, and Moody 1986, 1988; Loper, McCartney, and Buzyana 1988). Others, while agreeing with the significance of changes in mantle dynamics, see little evidence of underlying cyclicity (Vogt 1972, 1975, 1979; Courtillot and Besse 1987; Courtillot 1990; Larson 1991a, b). This chapter addresses the larger context of the geologic reorganization that occurred at the Paleozoic-Mesozoic boundary, the possible triggering of these events by mantle dynamics, and the controversy over cyclicity versus random events.

This debate should not obscure one critical point on which all of the protagonists agree: there is growing evidence of the connections between changes in the structure of the earth, tectonics, climate, and biologic diversity. One does not have to be an advocate of James Lovelock's Gaia hypothesis to appreciate the extent of these relationships, but considering these patterns provides an alluring framework in which to place our knowledge of the geologic changes across the Paleozoic-Mesozoic boundary. Yet it is essential to remember that the changes discussed in this chapter occurred over a far longer period than the end-Permian mass extinction. Thus, while they are not direct causes of the extinction, they are nonetheless important because the direct causes of the mass extinction, as discussed in the following chapter, may have been set in train by the processes discussed here. In essence, the question is whether the extinction was a chance occurrence or grew out of cycles within the mantle.

The material in this chapter is separated from the preceding chapters for two reasons. First, the model presented here represents attempts to synthesize the information presented earlier into coherent models of how the earth works—or might work. Although scientists (generally) require evidence to support hypotheses, we all have a tendency toward speculation of how we think the world works, and this chapter includes a number of such models. Unfortunately not all of these models are supported by as much data as ideas presented in previous chapters. So the second reason for placing them in a separate chapter is to emphasize that there is more speculation here than in earlier chapters. In general these models are consistent with our knowledge of the Permo-Triassic, but the data behind them may be a bit scanty. That really is not a drawback though, because only by imagining how the world *could* work can scientists generate models like these and stimulate others to go out and find data either confirming or refuting the models.

SUPERCYCLES IN EARTH HISTORY?

IN 1981 Alfred Fischer described what he called "two Phanerozoic supercycles." Cycles, and cycles upon cycles, are the essence of Fischer's view of the history of the earth. He sees in the alternation between two great 300-million-year cycles in mantle convection controls on sea level, continental movement, plutonism, and sedimentation, and through these on the amount of carbon dioxide in the atmosphere and thus climate and biotic diversity. Superimposed upon these long-term cycles are shorter cycles of about 30 million years (Fischer and Arthur 1977) and even shorter climatic cycles driven by astronomical perturbations with periods of 20,000 to 400,000 years.

Fischer's model of Phanerozoic supercycles begins with a period of rapid mantle convection leading to the breakup of supercontinents, an increase in volcanic activity, and a rise in sea level. The increased levels of CO_2 in the atmosphere combine with reduced weathering rates to generate a greenhouse climate. As mantle convection wanes, continental accretion begins, the volume of the ocean basins increase and sea level drops, volcanic activity slows, and weathering rates rise and atmospheric CO_2 is reduced, initiating a global icehouse state (figure 8.1).

Fischer marshals three lines of evidence in favor of the supercycle model. First, assuming that the volume of the oceans has been roughly constant, first-order eustatic curves should track changes in the volume of the ocean basins. As discussed in chapter 5, there are

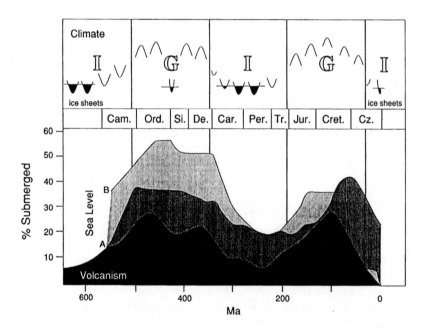

Figure 8.1. Fischer's two Phanerozoic Supercycles showing the climatic alternations between icehouse states (I) and greenhouse states (G). The blackened troughs within the icehouse cycles depict times of continental ice sheets. The lower portion shows sea level fluctuations. From (A) Vail, Mitchum, and Thompson (1977) and (B) Hallam (1977). Volcanism is based on granite emplacement in North America. From Engel and Engel (1964) and Fischer (1984).

three primary means of altering this value (see also Fischer 1984): 1) changes in the length of the spreading ridges (this length will be at a maximum during times with high numbers of plates and times of continental dispersion, and at a minimum during periods of continental aggregation); 2) changes in the rate of sea-floor spreading (faster spreading rates produce broad, high mid-ocean ridges and thus reduce the volume of the ocean basins); and 3) changes in continental thickness. Since the area of continental regions appears to be inversely proportional to their thickness, the breakup of a supercontinent decreases its mean thickness and should increase the volume of the ocean basins. In each case plate activity modifies sea level. Indeed, the pattern of first-order sea-level lows appears to correspond to intervals of continental aggregation and sea-level highs to continental dispersal, confirming that the eustatic curve is a proxy for levels of plate activity. Yet the correspondence is far from exact, and Fischer ascribes this lack of correspondence to lags between changes in continental movement and sea level.

A second index to plate activity is the changing patterns of volcanic activity. The temporal distribution of granite emplacement in North America broadly corresponds with first-order sea-level curves and with continental dispersion, in other words, with what Fischer construes to be peaks in plate activity (see also Johnson 1971).

The link to climate and biotic diversity is provided by shifts in the volume of carbon dioxide in the atmosphere. Variations in atmospheric CO_2 and the interplay between CO_2 and volcanism, global climate, and weathering rates have already been discussed. Fischer notes that as plate activity, and thus volcanism and sea level fall, the increase in land area will likewise raise the rate of oxidative weathering, resulting in the withdrawal of CO_2 from the atmosphere. Thus Fischer argues that atmospheric CO_2 levels were low during the late Precambrian, the late Paleozoic, and from the Oligocene to the present, intervals that correspond to increased glaciation. The intervening periods were times of increased CO_2 content and a resultant greenhouse effect (see also Berner 1989a, 1990, 1991 for a related model). In Fischer's model this alternation of the earth's climate between icehouse and greenhouse states is ultimately induced by changes in plate activity. The link to biotic diversity is somewhat more tenuous. Of the five major mass extinction episodes, only the end-Devonian event corresponds to a transformation from greenhouse to icehouse. The preceding chapters demonstrate that such a change occurred throughout the Permian and was largely responsible for long-term change in terrestrial plants and animals. Despite the overall significance of these climatic changes they occurred over a far longer time scale than the end-Permian mass extinction.

Fischer's supercycle model largely fits the available evidence, but as a description of pattern it lacks any discussion of process other than a vague claim for changing rates of mantle convection. Tom Worsley and his colleagues Damian Nance and Judith Moody have expanded this model considerably both in temporal scope and by providing a mechanism (Worsley, Nance, and Moody 1984, 1986; Worsley, Moody, and Nance, 1985; Nance, Worsley, and Moody 1986; Worsley and Nance 1989). Building on J. Tuzo Wilson's description of the episodic opening and closing of the Atlantic Ocean, the Wilson cycle, and Anderson's (1982) ideas that the insulating effects of supercontinents cause the development of asymmetric global heat flow patterns, Worsley and colleagues developed a model for "long-term quasi-periodic episodicity in plutonic, volcanic, orogenic, and metamorphic history" (Worsley, Nance, and Moody 1986:562). Their model cycle begins with the formation of a supercontinent, which may persist for up to 80 million years before thermal uplift caused by reduced heat flow through the supercontinent initiates rifting and eventually the breakup of the supercontinent. The resulting continents disperse for a maximum of about 160 million years before the oceanic crust between the dispersing continents becomes solid and dense enough for the onset of subduction within this Atlantic-type ocean. As subduction proceeds this ocean closes, forming another supercontinent and bringing one of these 440-to-500-million-year cycles to a close. In a fascinating intuitive leap, they suggest that the hemispherical asymmetry in the heat flow pattern (and consequent shape of the geoid) reflects a relatively permanent pattern:

> The ocean hemisphere (Panthalassa) tends to remain an area of general mantle upwelling, the net effect of which will be the tendency to maintain an antipodal supercontinent (Pangea) in an approximately stationary position. However, heat buildup beneath an assembled, stationary Pangea will cause an upward displacement of its geothermal gradient. Panthalassa will, therefore, remain the principal mantle heat dissipator. (Worsley, Moody, and Nance 1985:563)

Moreover, the model has been extended back to at least 2.6 billion years ago based on evidence of distinct orogenic episodes centered on 2.6 billion years ago, 2.1 billion years ago, 1.8–1.6 billion years ago, 1.1 billion years ago, 650 million years ago, and 250 million years ago, each of which is followed 100 million years later by apparent rifting events (Nance, Worsley, and Moody 1986, 1988). (Recent developments suggest that the supercontinent cycle does not extend into the Neoproterozoic, when the configuration of the continents was incompatible with this model [Dalziel 1992].)

Relying on calculations of the expected mean age of oceanic crust (and thus density, since older crust becomes colder and more dense) during different phases of the supercontinent cycle, Worsley and colleagues were able to construct a generalized sea-level curve through a complete supercontinent cycle. The explanatory power of Worsley's model is considerable, as shown in figure 8.2. The first set of graphs illustrates the correlation between the model sea-level curve (figure 8.2A) and global sea level as deduced by Vail's onlap-offlap model, the changing number of continents, and the percentage of the continents covered during transgressions. There is obviously a broad correlation between the expected changes based on the model and these various estimates of first-order sea-level change.

A second expectation of the model is that these tectonic changes should have a strong influence on climate and biotic diversity, changes that should appear in the fluctuations in the stable isotopic record of carbon, sulfur, and oxygen. Rifts associated with continental breakup are frequently sites of large-scale evaporite deposition. Similarly, Worsley and colleagues expect that when sea level is low, the increased erosion will deliver greater nutrients to the oceans, enhancing productivity and transferring light carbon to organisms. Therefore, carbonates, which record changes in the inorganic carbon reservoir, should have higher amounts of C^{13} and lower levels of C^{12}. The lower graph (figure 8.2C) shows exactly the patterns predicted. Additionally, Barley and Groves (1992) have suggested that supercontinent formation is correlated with peaks in the formation of metal deposits.

The link to Fischer's greenhouse-icehouse model comes from changes in CO_2 levels during the supercontinent cycle (see particularly Worsley and Nance 1989 and Worsley, Nance, and Moody 1986). The increased weathering rates during continental exposure should draw CO_2 out of the atmosphere, as discussed in chapter 5. This will tend to reduce greenhouse effects, and, if continental areas are near the poles, increase the likelihood of glaciation.

213

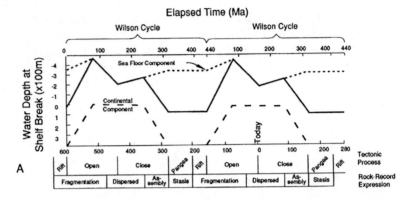

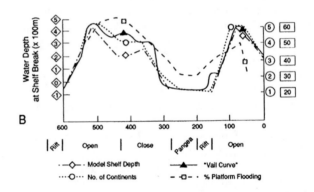

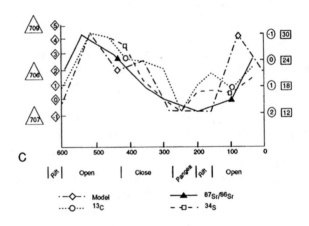

THE SUPERPLUME HYPOTHESIS

FISCHER AND WORSLEY and colleagues may well be correct about the link between mantle processes, expressed through changes in plate activity, and climate, but several geologists have argued the link is episodic rather than cyclical. For example, Vogt (1972, 1975, 1979) argued that the formation of mantle plumes are marked by increased frequency of magnetic reversal. Furthermore, Vogt claimed that mantle plumes like the one under the Hawaiian Islands were responsible for global cooling by continental uplift (increasing continentality) and by injection of volcanic dust. He also suggested that they might be linked to changes in oceanic circulation and sea level. Vogt regards increased rates of introduction of volcanogenic trace metals associated with the plume events as the most likely cause for the correlative extinction events. Hays and Pitman (1973), following the earlier suggestions of Valentine and Moores (1970) and Hallam (1971), confirmed that the mid-Cretaceous transgression was caused by a pulse of rapid sea-floor spreading. They further postulated that increased plate motions and resulting transgression should stabilize climate and increase diversity.

Mantle plumes are now widely accepted features of plate tectonics, but superplumes, similar structures on a far larger scale, are far less widely accepted. Roger Larson has compiled considerable evidence for a mid-Cretaceous mantle superplume and also suggested that a similar plume may have occurred during the Pennsylvanian-Permian (Larson 1991a, b; Larson and Olson 1991; see also Larson and Pitman 1972). Larson estimated the rate of production of new oceanic crust

Figure 8.2. The supercontinent cycle of Worsley and colleagues. (A) Schematic of two complete supercontinent assembly-fragmentation cycles with correlated effects on first-order sea-level curves (i.e., ignoring glacial effects, etc). The solid line shows the overall changes in water depth at the shelf break. The sea-level component (dotted line) shows the contribution to overall sea-level change from variation in the average age of oceanic crust. The continental component (dashed line) is the relative sea-level change from thermal doming of the supercontinent. (B) Comparison of the model first-order sea-level curves with the actual curve, the inferred number of continents, and the percentage of platform flooding (right-hand scale). (C) Comparison of the model (open diamond) with first-order changes in stable isotopes of carbon, sulfur, and strontium. From Worsley, Nance, and Moody 1986, after Worsley, Nance, and Moody 1984, and Worsley, Moody, and Nance 1985.

over the past 150 million years and identified a 50–75 percent increase from 120 to 80 million years ago. Since this pulse in activity appears largely in the Pacific oceanic ridges and plateaus, Larson suggests that it represents the eruption of a mantle superplume that developed at the core-mantle boundary about 125 million years ago. This interval coincides almost exactly with the Cretaceous Long Normal Super-chron (LNS), a 41-million-year period of stable magnetic polarity, which Larson relates to a modification of the temperature gradient within the mantle, an increase in convection in the outer core, and a corresponding decline in the frequency of magnetic reversals (Larson 1991a, b).

Caldeira and Rampino (1991) investigated the climatic effects of the CO_2 expected to be injected into the atmosphere during the eruption of the superplume. Previous work had suggested that mid-Cretaceous climates were 6° to 14°C warmer than today, largely due to increased sea level and the positions of the continents. Caldeira and Rampino's results suggest that as much as 2.8° to 7.7°C of the increase could have been due to increased atmospheric CO_2 from the super-plume. After considering other factors, particularly paleogeography, they conclude that no more than 20 percent of the total increase was due to volcanic CO_2.

Larson has also noted that increases in global temperature, deposition of black shales, oil formation, and sea level correspond to the apparent Cretaceous superplume and a possible second superplume during Pennsylvanian-Permian time (figure 8.3; Larson 1991a). Assuming long intervals of stable magnetic polarity stem from the formation of superplumes, Larson identifies the Kiaman Long Reversed Superchron (see chapter 5) as a similar interval. The late Paleozoic interval was a time of major coal deposition, which Larson likens to the oil formation of the mid-Cretaceous, and also, like the mid-Cretaceous, a period of increased gas production. Other broad similarities include increased sea level, but until the mid-Permian the Permo-Carboniferous glaciation produced a climate most unlike those of the mid-Cretaceous. Larson excuses this anomaly with the statement that even the increased CO_2 from a superplume was insufficient to moderate the ongoing global cooling. Clearly the case for a late Paleozoic superplume must be considered tentative, at best. What of the evidence for the mid-Cretaceous superplume?

The difficulty with Larson's hypothesis, as beguiling as it appears, lies in the connection he draws between oceanic crustal production

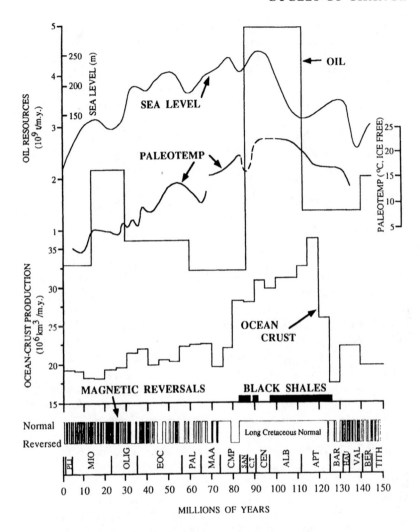

Figure 8.3. Correlations between magnetic reversals, global ocean crust production, high-latitude sea-surface paleotemperature, global sea level, black shale deposition, and global oil formation, suggesting a link between formation of mantle superplumes and other events. From Larson 1991a.

217

rates and deep-mantle convection. Larson assumes that subduction rates were stable during this interval, but if they were substantially greater this alone might account for the pattern he observes (Cox 1991). Determining rates of oceanic crust production along now-destroyed mid-ocean ridges is a very tricky business and there is far more uncertainty in Larson's estimates than he acknowledges (Griffin 1992).

A strikingly different view of mantle dynamics during the mid-Cretaceous suggests that reduction and then the elimination of magnetic reversals from 170 to 80 million years ago corresponds to a drop in instabilities in the upper core and a decrease in heat transfer from the core to the lower mantle. This would increase the D" layer immediately above the core-mantle boundary. This layer may have broken up near the end of the LNS (about 80 million years ago) and developed into one or more mantle plumes. The ascent of these mantle plumes reestablished mantle convection (Courtillot and Besse 1987; Courtillot 1990; Loper, McCartney, and Buzyana 1988). The eruption of these plumes had catastrophic effects, however, generating both explosive acidic volcanism and flood basalts (such as the Deccan Traps in India, dated at 65.7 ± 2.0 million years [Courtillot and Besse 1987]). Courtillot and Besse analyzed the rate of change in true polar wander (TPW, which is estimated from the movement of hot spots, which are assumed to be fixed relative to the mantle, against the spin axis of the earth). They identified an apparent decrease in the rate of TPW between 170 and 110 million years, preceding the Cretaceous LNS and coinciding with an increase in continental dispersal. They argue that TPW is a sensitive indicator of the rate of mantle convection, thus slow TPW and lower reversal rates indicate reduced convection.

Of greater interest here is the similarities Courtillot and Besse (1987) find between these patterns and those of the Permian. Like Larson, they see the tracks of a large mantle plume in the Kiaman Superchron but, unlike Larson, they argue that the plume erupted at the end of the superchron in the Siberian Traps, rather than at the outset of the interval of stable magnetic polarity (Courtillot 1990). Loper, McCartney, and Buzyana (1988) also see in the possible correlations between periodic mantle-plume eruptions and endogenous geologic events an antidote to extraterrestrial forcing mechanisms for periodic mass extinctions. This brings us back to an issue raised earlier: the possibility of periodic behavior in the formation of mantle plumes and their surficial expression in flood basalts and volcanism.

ARE FLOOD BASALTS CYCLIC?

RESOLVING THE age of the Siberian flood basalts, as discussed in chapter 5, is crucial to establishing the relationship between the extinction and the flood basalt, and Rampino and Strothers's (1988) correlation between mass extinctions and flood basalts over the past 250 million years. Rampino and Strothers (1988) established a correlation between the radiometric ages of 11 known continental flood basalts (table 8.1) from the Permian to the present. Most published dates for continental flood basalts are based on the traditional ^{40}K-^{40}Ar method, which may give dates that are too young if ^{40}Ar has leaked from the system, rather than newer, more precise techniques such as ^{40}Ar-^{39}Ar dating. The most accurate ages available for these 11 flood basalts suggest that each erupted in only a few million years. The flood basalts correlate well with mass extinctions previously identified by Raup and Sepkoski (1986). The Cenomanian mass extinction (91 million years ago) is the only post-Paleozoic mass extinction not associated with a continental flood basalt, but the extinction is roughly correlative with extensive oceanic flood basalts in the Western

TABLE 8.1 Apparent correlation between continental flood basalts (after Rampino and Strothers 1988) and mass extinctions (after Sepkoski 1990) over the past 250 million years. Note the Bajocian extinction, which appeared to correlate with the Antarctic flood basalt, has disappeared on further analysis, to be replaced by a very minor peak in the Callovian (Sepkoski 1990).

Flood Basalt	Age (Ma)	Mass Extinction	Age (Ma)
Columbia River (US)	17 ± 1	L-U Miocene	14 ± 3
Ethiopian	35 ± 2	Upper Eocene	36 ± 2
Brito-Arctic	62 ± 3		
Deccan (India)	66 ± 2	end-Cretaceous	65 ± 1
		Cenomanian	91 ± 1
Rajmahal (India)	110 ± 5	Aptian	110 ± 3
Serra Geral (S. America)	130 ± 5		
South-West African	135 ± 5	Tithonian	137 ± 7
Antarctic	170 ± 5		
South African	190 ± 5	Pliensbachian	191 ± 3
E. North American	200 ± 5	Norian	211 ± 8
Siberian	250 ± 2	end-Permian	250 ± 4

219

Pacific (Rampino and Strothers 1988). Rampino and Strothers performed a time-series analysis of the apparent initiation dates for the flood basalts, with the results suggesting a cyclicity of either 22 or 26 million years, or 31 million years, depending on whether or not the Siberian traps, Columbia River basalts (which may have a origin different than the other flood basalts), and the possible oceanic flood basalt about 90 million years ago are included. Time-series analysis is a common statistical technique used to establish whether or not periodic or cyclic behavior exists in a series of data through time. The periodicity uncovered by Rampino and Strothers is strikingly similar to the cycle in mass extinctions proposed by Raup and Sepkoski (1984, 1986a, b; Sepkoski 1989), and to a variety of other apparent geologic cycles (table 8.2). This raises the obvious issue of the causal link behind the correlation.

Loper and colleagues (Loper, McCartney, and Buzyana 1988) believe all of these events can be linked to the consequences of episodic generation of instabilities in the D" layer at the core-mantle boundary and the formation of large mantle plumes. One can envision a system that requires about 32 million years (the mean of the events in table 8.2) to thicken the D" layer sufficiently to initiate mantle plumes, thus establishing a quasi-periodic system.

The unavoidable conclusion from these models (other than the fact that more data would be nice) is that there is substantial evidence that the formation of Pangean-type supercontinents has consequences for endogenous geologic processes. Anderson's (1982) recog-

---

TABLE 8.2 Apparent periodicities in a variety of endogenous cycles. Note that only those cycles which actually correspond in period to the extinction cycle are of interest, and the extinction cycle of Raup and Sepkoski is about 26 myr. Hence none of these cycles will remain in phase with the extinction cycle.

Event	Apparent Period (myr)
Flood Basalts	34 ± 3
Alkalic and Silicic Volcanism	34 ± 3
Active Tectonism	33 ± 3
Marine Regressions	33 ± 3
Magnetic Reversals	30 ± 2

Compiled by Loper et al. (1988) from a variety of sources.

nition of the insulating effect of Pangea on heat flow and ensuing changes in the geoid, in the hot-spot pattern, and other tectonic events, seems relatively well-established. Likewise, the eruption of the Siberian Traps virtually coincident with the Permo-Triassic boundary as recognized in South China strengthens the arguments of those who discern a close connection between mantle dynamics, climatic change, and biotic diversity. Yet, once again, correlation does not constitute proof. There is obviously considerable disagreement among the various models discussed here over the sequence of events that leads to the initiation of mantle-plume activity, the connection between mantle plumes, and the frequency of magnetic reversals, and possible periodicity of mantle activity. The available geologic evidence for the Siberian Traps supports Larson's (1991a, b) models but not those of Courtillot and Besse (1987), or Loper, McCartney, and Buzyana (1988). Kent, Storey, and Saunders (1992) claim that crustal extension, basin formation, and uplift of the Siberian region began about 75 million years before the eruption of the Siberian Traps. Together with evidence from other continental flood basalt provinces, this suggests the head of the plume impacted the base of the continent during the Carboniferous, and a considerable period of plume incubation is required before eruption of a flood basalt. If Kent, Storey, and Saunders are correct, the incubation period for the plume beneath Siberia roughly corresponds with the duration of the Kiaman Superchron, data that match Larson's (1991a, b) model more closely than the alternative model, but which still requires the eruption of the plume from the D" layer prior to the beginning of the Kiaman Superchron.

By now it should be clear that almost all of these models combine events occurring on several different time scales and thus lack sufficient precision to generate testable hypotheses. Events separated by tens of millions of years can appear nearly simultaneous, and thus correlative, if the time dimension is sufficiently compressed in an illustration. A scientific model is only useful when it provides a new way of looking at the world, when it provides new insights into the way the earth works, or a series of hypotheses subject to testing either by additional simulations or empirical investigation, and preferably both. The models of Fischer and Worsley and his colleagues each provide attractive, perhaps beguiling, visions of how the world could have worked, but it is not clear that they offer more than "just-so stories." They suffer, in my view, from a failure to provide an agenda for future research.

The first step in generating such an agenda is to identify more precisely the questions to be addressed. In the main these concern the nature of core-mantle dynamics and the likely relationship between these processes and mantle plumes. Do mantle plumes actually arise from the core-mantle boundary? How reasonable are the models of Loper, McCartney, and Buzyana (1988) and others that identify instabilities in the D" layer as the source of these plumes? Finally, and most obviously, considerable attention must be given to the relationship between plume initiation and reversal frequency. Either Larson is correct and plumes erupt near the beginning of magnetic quiet intervals, or Courtillot and Besse and Loper and colleagues are correct in their conclusion that the mantle plumes erupt near the end of such intervals. Given the inability of geologists to directly analyze the core and mantle, many of these questions must be addressed through simulations.

The second area requiring further study is the relationship between climate change and both explosive pyroclastic volcanism and flood basalts. This subject has been an arena of intense investigation, but few definitive answers are yet available. A rigorous comparative approach is necessary in which all available events greater than a set threshold are investigated for their climatic and biologic effects. Frequently emphasis is placed on well-studied events, often near extinction episodes, and similar events without apparent climatic or biotic effects are either ignored or explained away with ad hoc arguments.

My guess is that the eruption of the flood basalts was probably closely tied to the insulating effect of Pangea and may turn out to have been a principal factor in the rapid end-Permian marine regression, the Kiaman Reversed Superchron, and indirectly to the end of the Paleozoic "icehouse." I have far less faith in any of the claims for periodicity, either over hundreds of millions of years or over some 30 million years. The models proposed by Fischer and Worsley and others, and the various mantle-plume hypotheses have an undeniable attraction for many in their unifying view of earth processes. Yet they remain unsatisfying somehow, in part because they set no agenda for further research, which in my view is the true test of a useful theory, and perhaps also because by explaining everything one wonders if, in the end, they have explained nothing. Leaving these more general theories behind, it is time now to turn to the possible causes of the end-Permian extinction.

IX

The Mother of Mass Extinctions

W E HAVE reached the point in our mystery where the detective gathers all the suspects together in the drawing room, recounts the crime, and, with a flourish, exposes the culprit. Previous chapters have touched upon several of the suggested causes of the end-Permian mass extinction. In this chapter we will examine the cast of characters advanced as explanations (figure 9.1) in light of the geologic, geophysical, climatologic, biostratigraphic, and paleontologic information discussed in the preceding nine chapters. And, at the conclusion of the chapter, I will unveil my thoughts on the cause of the end-Permian mass extinction. Unlike Hercule Poirot, Lord Peter Wimsey, or Sherlock Holmes, so many clues have not been uncovered or may have been misinterpreted that there is every likelihood that I am wrong in the suspect I finger. Science only progresses as a series of (we hope) closer approximations to the truth, so perhaps hidden in this examination of motive and opportunity are clues that someone else may use to find the killer.

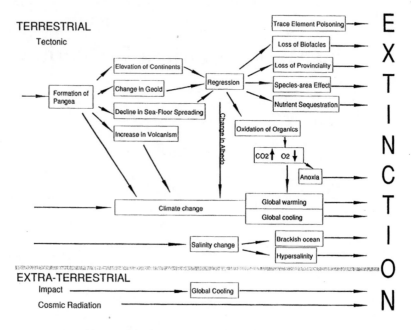

Figure 9.1. The possible direct causes of the end-Permian mass extinction are to the right, with the more indirect causes progressively to the left. Some events produce a number of secondary effects that may have contributed to the extinction.

In evaluating the proposed mechanisms several points should be kept in mind:

- Changes invoked as causes of the extinction must have occurred during the appropriate interval. There is little point in invoking change X as the cause of the extinction if it occurred 15 million years before the onset of the extinction. (Several widely cited hypotheses founder on this seemingly obvious point.)
- Present evidence suggests the extinctions within the marine realm were somewhat more severe than among terrestrial tetrapods, plants, or insects, although significant extinctions appear to have occurred in all three groups. The connnection between the terrestrial and marine events remains unclear, but this pattern constrains any discussion of mechanism.
- Cause and effect must be disentangled. This is frequently difficult but many of the mechanisms advanced as explanations are

based on patterns that may well reflect the effect of the extinction rather than its cause. Both threshold effects and positive and negative feedbacks may be significant. For example, salinity changes within a fixed range may have little effect on biotic diversity, but as salinity changes beyond the physiological tolerance of most species large portions of the biota may be decimated.

- Finally, many of us have a curious yearning for nice, simple explanations of complex historical events. It would be nice if there were a nice simple pattern of causality behind these events, and many people claim to find such simple explanations. They are almost always wrong.

CATASTROPHIC OR GRADUAL, OR, HOW GRADUAL IS GRADUAL?

PERHAPS THE most critical question in evaluating the various extinction mechanisms is the rate of extinction. Many earlier extinction papers considered diversity changes on the scale of entire periods (e.g., Newell 1962, 1967b) and advancing to stage-level resolution was a considerable improvement. While stratigraphic resolution may not have allowed greater precision in the past, low-resolution data encourage speculation on wholly inappropriate time scales. Analyzing rates of disappearance obviously requires high-resolution stratigraphic analysis, a dependable basis for intercontinental correlation, and good systematic work. The finer stratigraphic resolution achieved in recent studies makes it obvious that the events of the end-Permian mass extinction are limited to the close of the period. Thus, there is little point in considering events that occurred early in the period, or processes acting over longer time scales.

Most Western geologists who have studied the extinction in recent years concur that the extinction took place over a considerable span of time, likely beginning toward the close of the Guadalupian and accelerating during the Djulfian and Changxingian. Certainly the overall pattern of disappearances supports this view, with the caveat that normal marine communities hung on a bit longer in South China. However, this is the expected pattern even if the extinction were concentrated toward the end of the period. The smearing effect of the Signor-Lipps Effect (chapter 3) will be considerably enhanced by the marine regression. The many Lazarus taxa demonstrate the inadequacy of the Late Permian fossil record as a guide to marine diversity. Taken as read, the stable isotope record appears to support the view that the extinction was largely confined to the last several million years of the Permian, and largely coincident with the final pulse of the regression.

Enter the South China fossil record. Many respected Chinese geologists have argued for a catastrophic extinction at the close of the Changxingian and rejected a more gradual extinction as an artifact of poor preservation (see particularly Xu 1992). The discrepancy between Chinese and Western geologists may be due to one of three factors, as noted previously: first, the geographic isolation of South China may have created a refugium in which Permian marine communities persisted longer than elsewhere, a position perhaps supported by Sweet's (1992) graphic correlation results and equivocally by stable isotope studies. Second, the Signor-Lipps effect may have so smeared out the extinction pattern as to convert a catastrophic pattern into an apparently gradual pattern in much of the world, but fortuitously preserved a more accurate picture in South China. Finally, the apparently catastrophic pattern described by several Chinese geologists could be exaggerated by correlation problems within South China, poor taxonomy enhancing apparent diversity, or a major hiatus at the boundary. Missing sections at the top of the would explain the many abrupt range truncations. This position is supported by the abrupt shift in the carbon isotopic record. These three alternatives are not mutually exclusive, and all may be operating to some degree.

Clearly there is considerable work still to be done analyzing the patterns of extinction, but my best guess is that the extinction occurred over some 3 million years—and perhaps as many as 8 million years—and the pattern of decline was largely coincident with that of the marine regression (which does not mean, oddly enough, that the regression was necessarily the *direct* cause of the extinction). I suspect that the magnitude of the extinction in South China has been magnified by a number of problems, but that normal marine conditions did persist longer in South China than elsewhere. Whether this implies that the apparent Permo-Triassic boundary in South China is younger than that in western Tethys remains a topic for further research. Mechanisms operating over appreciably shorter or longer time spans than this are precluded, although they may have set the stage for other mechanisms.

A quick perusal of table 9.1 reveals a complex pattern of connections between the various extinction mechanisms. For example, a tectonically induced decline in sea-floor spreading may have been responsible for the marine regression, which in turn had a variety of possible effects on global climate (through changes in the earth's albedo), on atmospheric and ocean chemistry (through increased oxidation of carbon on newly exposed shelves), and on biofacies. Any one of these factors may have been directly responsible for the extinctions. Because of the various interactions between different hypotheses I have divided them into diversity-independent and diversity-dependent mechanisms (following Valentine 1973a) and within these categories from the least likely to the most likely scenarios. Each hypothesis will be presented together with the evidence in favor and then tested against all of the evidence available. Finally, I present my own scenario for the cause of the extinction and offer an agenda for further research at the conclusion of the chapter.

TABLE 9.1 Classification of extinction hypotheses according to whether they predominantly effect marine or non-marine settings, or both, and whether they should produce a rapid or gradual extinction. In several cases some effects are not known or not well understood. See text for details.

Suggested Causes	Marine	Non-marine	Rapid	Gradual
Nutrient Reduction	X			X
Decline in Provinciality	X			X
Trophic Resource Instability	X	?	?	X
Habitat Diversity	X			X
Ecosystem Collapse	X	?	?	X
Extra-terrestrial Impact	X	X	X	
Global Cooling	X	X		X
Salinity	X			X
Species-Area Effects	X			X
Oceanic Anoxia	X		X	
Atmospheric Anoxia	X	X	X	
Pyroclastic Volcanic Eruptions	X	X	X	
Flood Basalts	X	X	X	
Trace Element Poisoning	X		?	?

DIVERSITY-DEPENDENT HYPOTHESES

DIVERSITY-DEPENDENT ECOLOGICAL factors act to limit population growth as population size increases. In many circumstances food supply is a density-dependent factor: as population size expands, more and more food is consumed until further growth in the population is controlled by lack of food. By analogy, density-dependent extinction models involve the depletion of required environmental factors such as oxygen or habitat area (Valentine 1973a).

Origination Rates

Several authors have claimed that the drop in diversity was due to a failure of origination, rather than extinction. This actually is not a density-dependent extinction model, but a denial of mass extinction altogether. Hüssner (1983) suggested that the formation of Pangea and subsequent regression reduced the opportunities for allopatric speciation and thus origination rates (so there was not really an extinction at all). He claimed that extinction rates are not higher than during other intervals (which is demonstrably not true) but the stage-level resolution of the data he used (largely derived from various volumes of the *Treatise on Invertebrate Paleontology*) is too coarse to be useful. Hüssner was primarily concerned with countering claims for catastrophic extinction at the close of the Permian, but did not address the more likely scenario of a prolonged extinction over several million years. A similar argument was advanced by Rhodes (1967), who argued against a catastrophic extinction at the close of the Permian and questioned the magnitude of the extinction, based in large part on crinoid diversity patterns. Like Hüssner, Rhodes suggested that the drop in diversity reflected a decline in the origination rate during the Early Triassic. These two possibilities are easy to test by calculating the length of time required for background extinction to eliminate as many taxa as disappeared during the end-Permian extinction. Analysis of cohort survivorship curves demonstrates the end-Permian extinction was equivalent to about 85 million years of background extinction (Raup 1978).

Nutrient Reduction

Primary producers form the base of the food chain and variations in the abundance of marine phytoplankton may trigger severe repercussions throughout marine ecosystems (and connected terrestrial ecosystems as well—consider the plight of Chilean seabirds when El Niño causes a decline in marine nutrients and a crash in anchovy populations). A variety of scenarios involving a reduction in nutrient input and consequent mass extinction have been advanced. Bramlette (1965) linked the end-Cretaceous mass extinction to a decline in the input of terrestrial detritus and nutrients following a decrease in orogenic activity. In his model Late Cretaceous landscapes were essentially peneplaned and the sluggish streams of the time were incapable of transferring what nutrients did exist to the oceans. A reduction in upwelling activity exacerbated the effect by reducing nutrient supply from the deep sea. The coupled changes in strontium and sulfur isotopes provide some support to Bramlette's suggestion, although other difficulties remain. Enlarging upon Bramlette's hypothesis, Tappan (1968) noted the relationship between continental topography, climate, atmospheric circulation, and oceanic upwelling, and, in particular, proposed that oxygen levels may have declined as a result of a reduction in primary productivity. Later, Tappan (1970, 1982) asserted that the heavy extinction of suspension feeders during the end-Devonian, end-Permian, and end-Cretaceous mass extinction implicated changes in primary productivity as the principal cause through increased sequestration of organic material and other nutrients on land, thus starving the oceans of needed nutrients. Under Tappan's hypothesis, the increased biomass associated with the late Paleophytic floras, coupled with the formation of extensive late Paleozoic coals retained phosphorous, nitrogen, and organic material on land. Over time, nutrients in the ocean would gradually accumulate in the deep ocean or be buried in sediments. Upwelling normally returns nutrients from the deep ocean to surface waters, but failure of upwelling would further reduce nutrient availability. According to this model the end-Permian extinction is very gradual, and Tappan cites the selective removal of suspension feeders and large carnivores as support for this model. These models have received some support (McCammon 1969) but have generally been rejected. Lipps (1970), for example, noted that the oceans would have to be rendered essentially sterile for Tappan's model to work.

Moreover, there is little support for either continental denudation during the end-Permian or for nutrient sequestration on land. This does not necessarily invalidate her point about elimination of primary producers, a point to which we will return later in the chapter, but under Tappan's model nutrient sequestration would have reached a peak during development of extensive Carboniferous coal swamps, not during the Permian. Nor can the change from the Paleophytic flora to the Mesophytic flora be held responsible since this happened diachronously over tens of millions of years. Furthermore, exposure of the shelves during the regression and erosion of buried organic matter should have increased the delivery of nutrients to the oceans, and this at the height of the extinction!

Biogeography

The effects of the formation of Pangea on plant and animal distributions have long been a favorite explanation for the end-Permian mass extinction, despite the fact that Pangea had formed tens of millions of years prior to the extinction and persisted long into the postextinction recovery. In general these hypotheses fall into three classes: first, a decline in diversity as a result of a reduction in the number of marine provinces; second, the elimination of the broad continental shelves during the end-Permian regression via the so-called species-area effect; and third, increasing instability of trophic resources caused by the climatic effects of Pangea. These various models actually span the gap between density-dependent and density-independent causes but will be considered here.

These models were developed during the early 1970s in the initial enthusiasm over the discovery of plate tectonics. Not surprisingly, geologists' knowledge of the movement of the continents, the timing of collisions, and the intricacies of plate tectonics were far less developed than at present. It would be easy to criticize these ideas on the basis of the knowledge gained over the past 20 years, but unfair, for many of these ideas were important contributions at the time. Less defensible, however, is the continued perpetuation of easily refutable ideas in countless introductory textbooks.

Provinciality

Since most species occur only within a single marine province, one of the major controls on global diversity should be the number of marine provinces. Similar communities in different areas of a single marine province tend to have roughly similar community composition (at least for the more abundant species). Thus the species within a nearshore sandy-bottom community will tend to recur throughout a province but will differ between provinces. Provincial boundaries are largely determined by abrupt climatic transitions, most commonly in temperature. Thus the large number of modern marine provinces reflects the strong temperature gradient between the equators and the poles. Today there are about 31 marine provinces, if molluscs are used to define provincial boundaries (Valentine 1973a, b). Over longer time spans plate movements also control the number of provinces (provinciality) by combining and splitting continents and through the climatic effects of the position and size of continents. When the continents are dispersed endemism will increase and species diversity should climb. Conversely, aggregation of the continents will reduce endemism and species diversity should fall (Valentine 1969, 1973a, b; Valentine and Moores 1970).

This theory, largely developed by Jim Valentine and Eldredge Moores, was based on suggestions by several earlier workers. Valentine and Moores hypothesized that the formation of Pangea by the closure of the Uralian seaway in Late Permian times forced a reduction in sea-floor spreading. Since the depths of the ocean basins are a function of the age of oceanic crust, a reduction in the rate of seafloor spreading will allow the mean age of the oceanic crust to increase, increasing the size of the ocean basins. The volume of the mid-ocean ridge spreading centers will also decline. The net effect should be a regression. As noted earlier, the regression in turn increased continentality and altered climate patterns, which may have disrupted trophic resources. Specifically, increased seasonality in nearshore waters should occur as the continents assemble, along with an increase in nutrient supplies and increased competition as endemic biotas are forced into competition by a merging of formerly discrete provinces. Consequently, global diversity should be lowest/ when supercontinents exist.

The history of global diversity and continental drift supported this hypothesis in the early 1970s, but as discussed in chapter 6, we now know that Pangea was largely formed in the Early Permian and reassembled in the Late Triassic with the collision of various bits, including North and South China, to form Asia. There was no effect on either provinciality or global diversity during either episode. In fact, provinciality remained high into the Late Permian with 9 provinces and 6 potential provinces during the Guadalupian, only a slight decrease from the 12 provinces and 5 possible provinces during the Early Permian (Bambach 1990). Other workers have come to similar conclusions (see Erwin 1985; Schopf 1979). In their original paper, Valentine and Moores (1970) noted that high latitudinal temperature gradients may preserve high provinciality even in the face of continental accretion. Valentine (1973b) noted that the configuration of the continents, particularly of Tethys, allowed diversity to remain high even after the formation of Pangea. Indeed, these factors account for the persistence of high provinciality into the late Permian.

The suggested relationship between the end-Permian regression, a decline in sea-floor spreading, and a reduction in the volume of mid-ocean ridges caused by the formation of Pangea is more problematic. As discussed in chapter 5, the relationship between sea level and plate movements is considerably more complex than previously realized. Suffice it to say that the initial slow regression during the Late Permian may well be related to the formation of Pangea, albeit in a more complicated fashion than visualized by Valentine and Moores, but the final latest Permian drop in sea level appears to require a different mechanism.

Trophic Resources

The next link in the Valentine-Moores hypothesis is the association between climate, environmental stability, and trophic resources. Valentine and Moores (1972, 1973; Valentine 1971) argued that the more continental climates and higher seasonality along the marine shelves of larger continents or supercontinents (see chapter 6) will increase the instability of nutrients, primary productivity, and other trophic resources. Since resource availability is more uncertain in highly seasonal environments specialists will be at a disadvantage, while species with broad trophic and environmental tolerances will be favored. If this model is correct, continental collision should reduce species diversity. As with the preceding factors, subsequent research has demonstrated that the relationship between species diversity and environmental stability is far more complex than realized in the early 1970s. Nonetheless, the late Permian regression may well have exacerbated climatic instability, which, in concert with other factors, may have played a role in the extinction.

Habitat-Diversity Effects

When rapid regressions follow a period of maximal transgression, faunas adapted to extensive epicontinental habitats may become "perched" and unable to track falling sea level (Johnson 1974). The transgressions may allow the diversification of stenotypic organisms within epicontinental seas, and if the subsequent regression occurs too rapidly for migration and adaptation, the rapid reduction in the areal extent of epicontinental seas will leave these species trapped. Of course this only works if faunas within provinces are highly endemic, otherwise widely distributed species may escape extinction in other regions. In a test of this hypothesis using Pleistocene and Recent molluscs from California, Valentine and Jablonski (1991) confirmed that regressions have a far greater impact on these perched faunas than on other faunal elements. Given the magnitude of the change in end-Permian seas, it seems quite plausible that, if endemism was high, many more marine extinctions may have been associated with habitat destruction than with species-diversity effects.

DIVERSITY-INDEPENDENT HYPOTHESES

DENSITY-INDEPENDENT EXTINCTION models involve factors that affect all individuals of a species equally, independent of the number of species present. For example, the effects of poisoning or an increase in cosmic radiation do not depend on the number of individuals of a species, but influence all individuals uniformly. Most extinction mechanisms fall into this category, particularly those invoked as the explanation for the end-Permian mass extinction since it seems likely that the apparently more severe effects of these mechanisms are required to explain the breadth of the extinction.

Extraterrestrial Causes

Impact

The realization that anomalously high concentrations of platinum group elements, particularly iridium, at the Cretaceous-Tertiary boundary in Gubbio, Italy, may signal an extraterrestrial impact (Alvarez et al. 1980) triggered a profound change in geology. Prior to 1980 extraterrestrial mechanisms were rejected out of hand by most geologists as untestable speculation. Although some volcanoes emit iridium, the concentrations are far below the levels found at the Cretaceous-Tertiary boundary. However, platinum-group elements are far more abundant in extraterrestrial materials, so increased levels should signal an increase in extraterrestrial input, probably from the impact of a meteorite or comet. The discovery of a significant increase in iridium concentration at Cretaceous-Tertiary boundary sites around the world provided an empirical test of impact hypotheses (as well as creating a storm of controversy). Subsequent work triggered by the Alvarez hypothesis has largely confirmed the impact of at least one 10–30 km bolide at the end of the Cretaceous. The Alvarez hypothesis also led to investigations of many other boundary sections, including the Permo-Triassic boundary.

The first analysis of boundary clays in sections at Meishan (Zhejiang Province) and Wachapo Mountain (Guizhou Province) in South China revealed less than 0.5 parts per billion (ppb) iridium. The trace-element geochemistry suggested that the clays were of volcanic origin (Asaro et al. 1982). Nor was an iridium anomaly discovered at Permo-Triassic boundary sections in southern Armenia (Alekseev et al. 1983). However, a series of papers by Chinese geologists (Sun et al. 1984; Xu et al. 1985; Li et al. 1986, 1989; a comprehensive review may be found in Xu et al. 1989) suggested the presence of an iridium spike in boundary clays in South China. Sun et al. (1984) reported iridium anomalies of 8 ppb and 5 ppb from the Meishan boundary clay but reanalysis of the same samples at Los Alamos National Laboratory revealed 0.002 ppb and 0.024 ppb, respectively (Clark et al. 1986; Orth 1989). Chinese scientists have analyzed a number of additional sections and have also recovered microspherules (Xu et al. 1989; although their origin is debated, see Yin et al. 1989, 1992). Xu Dao-yi's group believe that impact of a stony achondritic meteorite (to account for the low iridium abundances) triggered severe environmental and ecological disturbances, leading to the mass extinction. While they admit that some volcanism occurred at the close of the Permian, they reject extensive volcanism as a cause of the mass extinction.

However, repeated attempts by western geologists to verify the Chinese reports have proven fruitless (Clark et al. 1986; Orth 1989; Orth, Attrep, and Quintana 1990; Zhou and Kyte 1988). There are a variety of possible explanations for this discrepancy, including a difference in interpretation of the level of iridium that is considered an anomaly, or analytical difficulties in the Chinese laboratories. A slight anomaly could indicate concentration by clays or biologic processes or a prolonged period of nondeposition, allowing buildup of the normal, background amounts of extraterrestrial material (the initial reason for Walter Alvarez's interest in iridium was to assay changes in depositional rates).

The GK-1 core in the Carnic Alps shows two small peaks in Ir (Attrep, Orth, and Quintana 1991) The lower peak coincides with the $\delta^{13}C$ minima at the top of the Tesero Horizon (*after* the Permo-Triassic boundary, which occurs about 40 cm above the base of the Tesero Horizon). The second occurs at the second $\delta^{13}C$ minima higher in the section, associated with pyrite formation. The authors note that while they cannot completely rule out an impact origin for the lower iridium anomaly, the lack of microspherules, or shocked quartz, makes an impact an unlikely source of the iridium. Furthermore, the ratios of cobalt/iridium, nickel/iridium, and chromium/iridium are orders of magnitude larger than those found in chondritic meteorites. The upper iridium anomaly is associated with a reducing environment and pyrite formation.

Incidentally, several geologists have suggested that an impact may have triggered the Siberian Traps. Rampino and Strothers (1988) and Alt, Sears, and Hyndman (1988) made similar suggestions. However, although the Siberian Traps are not tied to a hotspot track, they are associated with a failed oceanic rift (chapter 6). Geochemical analysis would aid in determining whether an impact origin for the traps is likely, since flood basalts generally exhibit a geochemistry consistent with a deep-mantle origin, not the shallow-mantle geochemistry expected from an impact-derived basalt flow. In addition, dynamic arguments suggest that the viscosity contrasts between the plume and the surrounding mantle necessary for mantle-plume formation are generated near the core-mantle boundary (Loper and McCartney 1990). At present any connection between the Siberian Traps and an impact appears unlikely.

There is no positive evidence for an impact at the Permo-Triassic boundary, but since one can never prove that something *didn't* happen, the possibility still exists.

Cosmic Radiation

An increase in cosmic radiation from a nearby supernova was advanced by Schindewolf (1954, 1963) as an explanation of the Permo-Triassic extinction. Schindewolf believed that the lack of an erosional break at the Permo-Triassic boundary eliminated a terrestrial cause, leading Newell (1962:606) to comment: "The apparent lack of erosional hiatus at the era boundaries is interpreted by Schindewolf as suggesting that the stratigraphic record is complete even though there are clear-cut paleontological interruptions. . . . In other words, failure to discover physical evidence that an unconformity exists is taken . . . as proof that one does not exist." Newell went on to discuss many examples of paraconformities—sections that were seemingly continuous but actually missing millions or tens of millions of years of time (see also Newell 1967a).

Schindewolf was primarily interested in large-scale evolutionary patterns and was convinced that major innovations in the history of life (traditionally recognized as the appearance of new phyla, classes, and orders) were caused by an increase in the rate of mutation. Moreover, Schindewolf knew, as did everyone else in the 1950s, that radiation was a major source of mutation. Confronted with the multitude of new taxa that arose during the Triassic, Schindewolf concluded that an increase in cosmic radiation would have eliminated many groups and increased the rate of mutation among the survivors, thus explaining both the extinction and the subsequent radiation (see also Newell 1956, 1962).

More recently, Hatfield and Camp (1970) noted the rough correlation between mass extinctions and the position of the solar system relative to the galactic plane. They suggested that major magnetic fields (and coincident cosmic radiation) occurred within the galactic plane. The movement of the earth through this plane (with a cycle of 80–90 million years) would subject the earth to considerable magnetic fields and increased cosmic radiation. However, the period of increased exposure is considerably longer than the duration of the extinction, and this model provides no explanation for the concentration of the extinction during a comparatively brief interval.

A more telling objection to both models is that cosmic radiation will affect terrestrial and very shallow marine organisms more than benthic marine organisms. Both models are largely ad hoc and are contradicted by the available evidence. An increase in cosmic radiation might kill much of the terrestrial flora and decrease the supply of nutrients to the oceans, indirectly causing a mass extinction, a possibility hinted at by Bramlette (1965).

Global Cooling

The possibility that a brief episode of global cooling, perhaps involving bipolar high-latitude glaciation, triggered the extinction was discussed in chapter 6. To briefly reiterate, Stanley argued that glacial deposits on the Kolyama block in present-day Siberia and in eastern Australia, together with the disappearance of reefs, a cessation of carbonate deposition, and the depauperate faunas of the earliest Triassic, indicated an interval of global cooling (Stanley 1984, 1988a, b). Upon closer examination the glacial deposits turn out to be of mid-Permian age and apparently represent a final, bipolar pulse of glaciation at the close of the Permo-Carboniferous glaciation. There is no evidence for global cooling at the close of the Permian; indeed, there is considerably more evidence for warm climates but with a high temperature gradient between the poles and the equator.

Global Warming

Waterhouse (1973) and Dickins (1983) have suggested global warming as the cause of the extinction, although with little impact. Maxwell (1989) dismisses the idea, noting that tropical organisms should be able to escape extinction simply by moving toward the poles. At any rate these scenarios provide no mechanism for the global warming and are thus more in the realm of speculation than a testable hypothesis. Yet several of the mechanisms discussed below do involve greenhouse effects and global warming, so we will return to this possibility later in the chapter.

Salinity

The decimation of a variety of groups whose modern representatives are largely stenohaline (intolerant of salinity changes), including bryozoans, ostracodes, and corals, together with the minor effects in more euryhaline groups such as gastropods, led to suggestions that reduced salinity (brackish conditions) were responsible for the mass extinction (Beurlen 1956; Fischer 1965; Stevens 1977; Lantzy, Dacey, and MacKenzie 1977; Posenato 1991). Beurlen suggested that a gradual reduction in marine salinity during the Late Permian eliminated stenohaline marine taxa and that Late Permian faunas were largely derived from brackish-water lagoons and estuaries. In this model Lazarus taxa persisted in refugia with normal marine salinity and reappeared with the return to normal salinity (see discussion in Maxwell 1989). Stevens (1977) confirmed that groups which are today stenohaline, or which can be inferred to have been stenohaline during the Permian, suffered more severe extinction than other groups. Distribution patterns and reproduction limits of modern organisms suggest that a salinity drop from 34 to 36 parts per thousand (ppt) to 28 to 30 ppt would be sufficient to cause an extinction of the same magnitude as the end-Permian event (Fischer 1965; Stevens 1977). Is such a drop plausible? A drop of this magnitude would require removal of 7×10^{15} tons of salt (Fischer 1965). Stevens (1977) conservatively estimated the total volume of Permian salt at 1.6×10^6 km^3, or 10 percent of the present volume of salt in the ocean. He suggests that formation of this amount of salt would lower salinity to 31.5 ppt, or half the amount necessary to explain the extinction. Since these estimates certainly underestimate the total volume of salt formed, Stevens argues that correcting for the underestimate gives about the right drop in salinity.

Beurlen suggested variations in the rate of erosion of salt on land as the cause of the salinity fluctuations, a mechanism Fischer (1965:569) accurately described as "beginning with a dubious assumption and ending with a doubtful conclusion." Fischer proposed that extensive evaporite formation caused the formation of dense brines that flowed into the deep sea, reducing the salinity of the surface waters. Since the resulting evaporites would not have been subject to erosion (over the near term), they would have been effectively sequestered from further involvement in marine salinity. Fischer makes some back-of-the-envelope calculations suggesting that perhaps 30 percent of the total amount of salt in the oceans could have been sequestered by this mechanism from the Guadalupian through the end of the Permian. In addition to the drop in salinity the formation of deep-sea brines also would have eliminated upwelling, reducing the flow of nutrients to surface waters and interfering with oceanic circulation to the detriment of world climate (see also Thierstein and Berger 1978).

Benson (1984) pointed out that the brackish-water model of Fischer was a global version of what happened on a regional scale as the Mediterranean dried up during the Miocene. He pointed out that the ostracodes that survived this Messinian crisis were many of the same lineages that survived the end-Permian mass extinction, or were closely related. Yet after comparing the two events Benson concludes that the reduction in salinity postulated by Fischer, even if it had occurred, would have been insufficient to cause an extinction of the magnitude observed at the end of the Permian.

An opposing model was advanced by Bowen (1968), who argued that Permian climates triggered an increase in Permian salinity of about 20 percent above the present-day levels, based on the volume of massive Louann salt deposits from the Gulf Coast and other post-Paleozoic evaporites. He suggested that extensive Permo-Triassic evaporite deposition returned marine salinities to normal, but only after killing off most marine organisms.

These salinity hypotheses are instructive examples of how often "explanations" are nothing of the kind. Stenohaline taxa are also largely stenotopic, and independent evidence must be advanced that salinity changes were the selective factor. Contrary to several of these papers, nautiloids did not particularly suffer during the extinction, blastoids and crinoids disappeared long before the ammonoids or the brachiopods, and "strophomenid" brachiopods suffered far greater extinction than did spiriferid brachiopods. None of these patterns are consistent with the salinity hypothesis. Moreover, this model implicitly assumes that all Permian evaporites formed during the Late Permian, coincident with the extinction. Yet, as noted in chapter 6, the bulk of Permian evaporites are now believed to have formed during the early Late Permian, well before the onset of the extinction. Furthermore, analysis of fluid inclusions and patterns of evaporation suggest that the maximum likely salinity drop during the Permian was 5 percent, far less than necessary to explain the extinction. In summary, none of these patterns is consistent with the salinity hypothesis.

Species-Area Effects

The coincidence between marine regressions and the major mass extinctions (Jablonski 1986c) has led many paleontologists to suspect a connection between the two. Norman Newell has long argued that the regression was the major cause of the extinction (1962, 1967a, b, 1973, 1982) but the nature of the connection has proven elusive. One of the simplest connections is derived from MacArthur and Wilson's theory of island biogeography (1967). They suggested that species diversity on an island is a function of immigration to the island from a continental source, and extinction on the island due largely to competition. Thus the immigration rate should be a declining function of the number of species on the island and should approach zero when all the species from the source pool have reached the island. Similarly, as species diversity increases, the extinction rate should climb as competition for resources increases. The equilibrium species diversity will be the point where the immigration rate and the extinction rate are the same. Among the implications of the theory are that smaller islands and more distant islands should have fewer species than larger islands or those closer to the source area. This relationship provided a theoretical underpinning to many empirical observations, but has also met with considerable criticism. This relationship was soon placed in an evolutionary context with speciation replacing immigration.

If the number of species is plotted against area on a log-log plot a straight line generally results. The equation for this line is generally given as:

$$S = kA^z$$

where S is the number of species, A is the area in question, k is the intercept of the line, and z is the slope of the line. Empirical studies suggested that z = 0.26 for a wide number of taxa (Simberloff 1974). The species-area effect was employed to relate changes in the area of shallow marine seas to species diversity, and particularly to the end-Permian mass extinction (Schopf 1974; Simberloff 1974). This assumed not only the validity of the species-area relationship but also that reduction in shelf area through a regression would cause extinction as a new, lower equilibrium diversity was achieved. Using Schopf's (1974) estimates for changes in diversity, coverage of the continental shelves and duration, Simberloff (1974) plotted the number of families (rather than species) against area. He confirmed Schopf's claim that the magnitude of the reduction in shelf area would produce a decline in marine diversity consistent with the end-Permian extinction.

A variety of biologists and paleontologists have criticized the species-area effect (Conner and McCoy 1979; Flessa and Sepkoski 1978; Wise and Schopf 1981; Schopf 1979; Jablonski 1980, 1985; Hallam 1989a; Williamson 1988; Valentine and Jablonski 1991). In general, it does not appear that the relationship between species diversity and area is as tightly constrained as required by the hypothesis. Williamson (1988) questioned the entire basis of the species-area effect, concluding that sampling artifacts will produce a relationship similar to the species-area effect in small areas while increased habitat diversity is likely responsible for the effect in larger regions. Certainly recent paleontological studies cast doubt on the general applicability of the species-area effect. Schopf (1979) largely repudiated his earlier work, arguing that the number of marine provinces was a primary control on species diversity. In another study (Hansen 1987), the species-area relationship suggests a 50 percent reduction of shelf area along the Gulf Coast during the Middle Eocene should have reduced total molluscan species diversity from 384 species to 310. In fact, diversity was unchanged. Wise and Schopf (1981) also rejected the species-area effect based on their analysis of the Pleistocene sea-level fluctuations. Their analysis shows that a 200 m change in sea level would produce a 25 percent decline in species diversity, 10 percent change in genera, and only 5 percent in families. They concluded that changes in the number of marine provinces are a more effective control on diversity than the species-area effect, but it is not clear that their study is particularly applicable to the Permo-Triassic. Pleistocene sea-level changes were largely vertical and involved little change in the size of shelf seas, a situation quite unlike the Permian (Jablonski 1985).

If the species-area relationship is valid, regressions should have a far greater effect on continents than on islands since, in general, the area of an island will increase during a regression (Stanley 1979). Modern tropical reef biotas are among the richest environments in the world, rivaling if not surpassing the tremendous diversity found in tropical rain forests. If most marine families have representatives on oceanic islands they will be relatively immune to regression-induced extinction. Jablonski (1985; see also Jablonski and Flessa 1986) tested this expectation by surveying the diversity of 276 families of shallow marine bivalves, gastropods, echinoderms (asteroids, ophiuroids, and echinoids), and scleractinian corals on 22 oceanic islands. Each of these groups has remarkably high representation on these islands. Of the 276 families 239 (87 percent) are present on one or more islands and 220 (78 percent) on two or more. All 21 scleractinian families occur on at least one island, as do 97 percent of the gastropods, 82 percent of the asteroids, 80 percent of the bivalves, and 76 percent of the echinoid and ophiuroid families. Jablonski concluded that a marine regression that eliminated all families on continental shelves would cause a global decline of only 13 percent. Since total devastation of the continental shelf fauna is implausible and since there are actually thousands of oceanic islands, Jablonski concluded that the species-area effect was not a plausible explanation for the end-Permian extinction.

Jablonski carefully noted that his results are dependent upon several assumptions, in particular that Permian biotas were not highly endemic, that the species/family ratio does not invalidate the results, and that the larval dispersal strategies were not greatly different. The last assumption was discussed in chapter 3, where I pointed out that crinoids, articulate brachiopods, and gastropods may have all experienced the preferential elimination of nonplanktotrophic lineages during the end-Permian extinction. If this is confirmed, it might partially negate Jablonski's argument since lineages with planktotrophic larvae are more likely to be found on oceanic islands. On the other hand, however, many nonplanktotrophic families have wide geographic distribution today. Both bivalves and echinoids have significantly fewer species per family on oceanic islands than on continental shelves, while gastropods show no difference (Jablonski and Flessa 1986). This suggests that families within some clades might be more extinction-prone than others, but further work is needed to investigate the potential magnitude of the effect.

Jablonski (1985) visualized reefs and oceanic islands as simple conical structures. He neglected the distinction between the outer-reef slope and the portion of the reef inside the reef crest, including lagoons and barrier reefs. These inner-reef habitats are eliminated during regressions, yet they contain the bulk of the shallow-water habitat in the Pacific; only the outer-reef slope environments are preserved during regressions (Paulay 1990). Bivalve species restricted to inner reefs were more susceptible to extinction during the Pleistocene sea-level fluctuations than those inhabiting outer reefs. About 30–40 percent of bivalve species in inner-reef habitats suffered extinction. Topography is also an important control. The gently sloping reefs of the western Pacific experienced far less extinction than did islands in the central Pacific where reef crests tend to be well developed. Thus the effect of marine regressions through loss of specific habitats may be more important than the loss of area, both on continental shelves and on oceanic islands.

Anoxia-Stagnant Ocean

One of the most effective density-independent extinction mechanisms is simply lowering the oxygen content of the atmosphere. Most organisms are well adapted to the ambient atmospheric oxygen levels and fare poorly when confronted with anoxic environments. The genesis of these models was the proposal that the extensive black shales of the Paleozoic indicate that oceans were initially dysaerobic (Berry and Wilde 1978; Wilde and Berry 1984, 1986; see also Southam, Peterson, and Brass 1982). Modern oceans generally have an oxygen-minimum zone at mid-depths, with increasing oxygen at greater depths. The Berry-Wilde hypothesis replaces this with an anoxic deep ocean. Episodic expansion of this anoxic layer during glacial episodes may explain at least some Paleozoic mass extinction episodes, but does not apply to the post-Paleozoic. While this model does not appear to be directly applicable to the end-Permian, the isotopic record of carbon, oxygen, and sulfur discussed in chapter 8 and the various potential explanations for the isotopic shifts have generated a variety of hypotheses focusing on global anoxia and associated global warming as a cause of the end-Permian mass extinction (Keith 1982; Hallam 1989a; Wignall and Hallam 1992; Gruszczyński et al. 1989, 1990; Holser et al. 1991; Hoffman 1989d; Hoffman, Gruszczyński, and Małkowski 1990, 1992; Małkowski et al. 1989).

The first of these explanations (figure 9.2) proposed that exposure and oxidation of organic material on the continental shelves by the latest Permian regression led to a drop in $\delta^{13}C$ and exposure and oxidation of large amounts of organic matter on the continental shelves. This caused a drop in $\delta^{13}C$, reflecting an increase in the amount of CO_2 in the atmosphere (pCO_2), which in turn set in motion a series of other events, including an increase in surface temperature of about

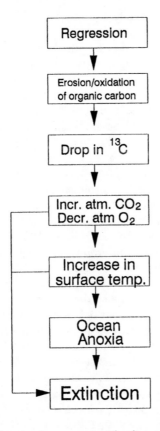

Figure 9.2. Postulated extinction scenario. The direct cause of the extinction is an increase in surface temperature, oceanic anoxia, and possible remobilization of metals. Modified from Holser et al. (1991).

6°C (reflected in the fall in $\delta^{18}O$). The drop in the solubility of oxygen in seawater, already depleted in oxygen as a result of long-term oxidation of carbon, would exacerbate oceanic anoxia (Holser et al. 1991). Specifically, the mid-depth oxygen-minimum zone that persisted throughout much of the Paleozoic would increase in thickness and likely invade shallow depths (Berry and Wilde 1978; Wilde and Berry 1984, 1986). Erosion of recent volcanic material, coupled with the incursion of anoxic waters, may have been responsible for the precipitation of metals in the western Tethys, as recorded well above the boundary in the GK-1 core. The major constraint on this model is the subsequent discovery that isotopic fractionation between carbonate and organic material did not change across the boundary, limiting the magnitude of the possible buildup of CO_2 (Magaritz et al. forthcoming).

Two other models involving anoxia have been advanced as well. Hallam (1989a, 1991; Wignall and Hallam 1992) reversed the usual perspective, proposing that the extinctions were not caused by the regressions but rather by the spread of anoxic waters during the subsequent rapid transgression (figure 9.3). I have previously discussed the rapidity of the early Triassic transgression, which may well have been the most rapid of the Phanerozoic (Holser and Magaritz 1987). This particular model is rapid, if not catastrophic, and largely restricted to the marine realm, since Hallam invokes no corresponding changes in atmospheric composition. It may be worth noting that this hypothesis is of a piece with Hallam's general view that many mass extinction events are a consequence of oceanic anoxia associated with rapid transgression (Hallam 1989a), notwithstanding the inconvenient fact that numerous Cretaceous anoxia events had no discernable effect on diversity (Kauffman 1992). It is true, however, that the Cretaceous anoxia events were not associated with the rapid transgression envisioned by Hallam.

Hallam marshals four lines of evidence in favor of this hypothesis. Earliest Triassic communities are low in diversity, although species abundance may be very high, characteristic of opportunistic expansion in a environmentally stressed setting, which Hallam and Wignall suggest was dysaerobic. They finger *Claraia*, the widespread early Triassic bivalve, as one of these dysaerobic elements. Less clear is why *Claraia* and its other latest Permian relatives are widespread *prior* to the extinction (Xu et al. 1989). Perhaps they were getting ready. Laminated black shales, diagnostic of anoxic conditions, are largely lacking, with the exception of a black claystone at the boundary in South

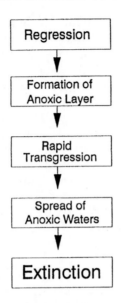

Figure 9.3. Suggested oceanic anoxia hypothesis, advanced by Hallam (1989a, 1991) and Wignall and Hallam (1992), involving the advance of an oceanic anoxia layer with the earliest Triassic marine transgression.

China (Li et al. 1991; Hallam 1989a). Unfossiliferous finely laminated limestones and clastics are far more common in the earliest Triassic. Wignall and Hallam (1992) examined Permo-Triassic boundary sections in northern Italy and in the western United States. They conclude from stratigraphic evidence that a minor early Dorashamian regression was followed by a late Dorashamian-Griesbachian transgression and the boundary, and in their eyes the extinction, occurred during the most rapid phase of the transgression. Needless to say, this model disagrees in almost every aspect with the standard interpretation of the upper Bellerophon Formation and the overlying Mazzin Member (which does not mean it is wrong—there is clearly considerable room for further study). Additionally, the ratio of carbon to sulfur is frequently taken as an index of anoxia, and Hallam (1991) uses Berner and Raiswell's (1983) C/S curve to argue that a minimum reached in the Early Triassic was indicative of global anoxia (see also Keith 1982). Finally, Hallam also invokes the Late Permian-Early Triassic cerium anomaly (see chapter 7) as an indication of anoxia.

While some level of anoxia may well have occurred during this interval, there are several reasons to approach Hallam's hypothesis with caution. Perhaps most telling is simply to note that all available evidence suggests that the biota was decimated prior to the onset of the transgression. Faunal ranges are not truncated by the onrushing seas but disappear well before the boundary. Presumably blastoids did not disappear at the close of the Guadalupian from stress over anoxic oceans 5 million years in the future. Cause and effect appear to me to be reversed in Hallam's scenario. The planar laminated beds of the Early Triassic signify the lack of bioturbation, but not necessarily because the waters are anoxic (although that is one possibility), but because most everything was dead. Hallam admits that black shales, the diagnostic feature of other well-studied anoxic events, are largely absent from Permo-Triassic boundary sections. In fact, in South China Bottjer, Droser, and Wang (1988) found evidence for exactly the opposite conditions described by Wignall and Hallam (1992). Bottjer and colleagues described redox cycle variation (alternating well-oxygenated sediments and dysaerobic or anaerobic sediments) in an outer-shelf environment below the boundary. Immediately above the boundary the sediments show extensive bioturbation, characteristic of well-oxygenated conditions. As far as the geochemical evidence is concerned, the change in the C/S ratio is based on a particular model of changes in the carbon cycle during the Phanerozoic. The C/S ratio is poorly constrained and employing it in support of this hypothesis is hardly credible. Likewise, the timing of the cerium anomaly is too poorly constrained to be informative. Hallam has made a number of important observations, but the coarse temporal resolution of his geochemical data, particularly when compared to the high-resolution stratigraphic studies of Holser, Magaritz, and colleagues, inspires little confidence in his hypothesis.

The third model seeking to explain the isotopic shifts at the boundary begins with a different view of the magnitude of the shifts themselves. In figure 7.3, based on analysis of sections of the Kapp Starostin Formation in west Spitsbergen, Gruszczyński et al. (1989) show $\delta^{13}C$ reaching a high of about $+7‰$ in the lower Tatarian and then falling to $-3‰$—a shift of 10 per mil, far greater than that seen at any other section. Using the same methods as in table 7.1, Makowski et al. (1989) argue that a shift of this magnitude is not possible under Holser et al.'s (1991) scenario. In a series of papers this group has developed an alternative model to explain the shifts between the carbonate and organic carbon reservoirs that relies upon oxidation of organic carbon sequestered in the deep ocean, rather than from shallow shelves (Gruszczyński et al. 1989, 1992; Hoffman 1989; Hoffman,

Gruszczyński, and Małkowski 1990, 1991; Małkowski et al. 1989; an earlier version of this model was proposed by Keith 1982). They see the spike in $\delta^{13}C$ to $+8‰$ as an indication of increased deep-sea sequestration of carbon within a stratified and stagnant ocean. Under such conditions, the ocean can be considered as two boxes, the upper, shallow-water box of well-oxygenated waters, and a lower, far larger portion that is largely anoxic. The two boxes are divided by the redoxcline, and have minimal interchange between them. Other characteristics of this system, which they call an "overfed ocean," include organic carbon deposition below the redoxcline, eutrophic conditions and enhanced nutrient regeneration, abundant oxygen production, and warm climates. This can be contrasted with "hungry oceans," where nutrients, particularly nitrogen and phosphorous, limit biomass production. These oceans are characterized by vigorous circulation through a well-oxygenated, homogeneous ocean (i.e., no division into two systems), recycling of organic carbon, deposition of nutrients, consumption of oxygen, and colder climates. The nutrient depletion in hungry oceans makes them capable of sustaining a far smaller biosphere than an overfed ocean. The similarities to Fishcer's greenhouse-icehouse model discussed in chapter 8 should be obvious.

Małkowski, Gruszczyński, and Hoffman (1991) suggest there are a number of sedimentary indicators that should be associated with these models. Overfed oceans should have enhanced carbonate deposition above the redoxcline and dissolution of silicious deposits below the redoxcline. In contrast, on a global scale carbonates can be subject to dissolution and silicious deposits to accumulation in a hungry ocean. In this scenario this geochemical change is primarily responsible for the absence of carbonate deposition in the Early Triassic. Under this model the transition from a stagnant to a mixed ocean was also responsible for the sudden shift in strontium isotopes as discussed in chapter 5 (Gruszczyński et al. 1992).

These authors argue that the transition between these two states occurred at the Permo-Triassic boundary, involving rapid destratification of the oceans and establishment of vigorous circulation. This resulted in oxidation of the deep-ocean carbon that had been sequestered during the previous interval. The oxidation is marked by the decline in $\delta^{13}C$ and $\delta^{18}O$ and the increase in $\delta^{34}S$. More importantly, however, the oxidation removed both phosphorous and nitrogen in the marine realm and induced extinction through nutrient deficiencies. On land, extinctions were caused by a drop in atmospheric oxygen of between 10 and 90 percent and probable climatic cooling.

The hungry ocean-overfed ocean scenario appears to be an equally plausible explanation of the isotopic shifts as the erosion and oxidation of organic matter from the continental shelves. More importantly, how well does this model match the available data? Here several problems surface. First, the model is dependent upon the rapid increase in $\delta^{13}C$, followed by an abrupt drop toward the boundary, yet Spitsbergen section is the only one examined to date that shows this pattern. Granted the microsampling they conducted is the most precise approach available, but the lack of corroboration from other studies and the general similarity between other sections is troubling. Second, the constraints on the sampling of the Spitsbergen sections are rather loose. Several sections were sampled and combined into a single composite section and the little information is given on the biostratigraphic resolution. Certainly neither the sampling scheme nor the biostratigraphic zonation approach the resolution of the GK-1 core (Holser et al. 1991). Third, Hoffman and colleagues find support in the pattern of carbonate and sediment deposition for their model. However, the pattern of chert deposition contradicts their model. Bedded chert and silicified fossil assemblages are common through the Permian up to the Upper Changxingian in South China. Both chert and silicified deposits disappear at the boundary, however, and appear to be largely absent from the Early Triassic. Fourth, there is no evidence that the biota of the late Paleozoic or the Late Permian were any more abundant than they are today, and substantial evidence that far less biomass existed, again contrary to the pattern proposed. Finally, no mechanism for the rapid shift between the overfed and hungry states has been advanced, although one could invoke volcanogenic/hydrothermal heating (Vogt 1989), or perhaps the eruption of a superplume, as occurred during the mid-Cretaceous. In summary, the lack of the support for the magnitude of the isotopic shift from the Spitsbergen section renders this model an unnecessary explanation of the isotopic data. The problems noted above suggest that oxidation of organic carbon from the continental shelves and methane from the continental shelves and slopes is a more parsimonious explanation of the isotopic data.

Volcanism

The obvious climatic effects of many of the comparatively small volcanic eruptions during historic times have fueled considerable speculation about the impact of the far larger events of the past. There are four possible mechanisms through which massive volcanism might be able to cause mass extinctions: creation of a dust cloud that reduces photosynthesis and initiates global cooling; injection of massive amounts of CO_2 and sulfates into the atmosphere causing global warming, creating acid rain as the sulfate is converted to sulfuric acid and reducing the protective ozone shield; creation of a thermal anomaly; and injection of poisonous trace elements into the atmosphere and oceans. Volcanic eruptions affect climate in two ways, and over two different time spans. First, injection of dust and ash into the stratosphere may reduce incoming solar radiation sufficiently to cool the earth as well as produce pretty sunsets (see review by Rampino, Self, and Strothers 1988). The Tambora eruption in 1815 and Krakatau in 1883 provide our primary experience with volcanically induced climate change. Tambora cooled global climate by about 1.5–2.0°C for about three years. However, dust particles rapidly aggregate into larger particles and settle out of the atmosphere. It appears to be impossible to maintain a dust cloud in the atmosphere for longer than 3–6 months and there is substantial doubt that the dust cloud from even a massive eruption would be sufficient to cause global extinction.

The second climatic effect is more significant, and more long-lasting as well. Volcanic eruptions introduce enormous quantities of aerosols into the atmosphere, particularly CO_2, SO_2, and H_2S, and since sulfate aerosols persist longer in the stratosphere than silicate dust particles, they should have a substantial cooling influence over a geologically significant interval. The eruption of Mt. St. Helens in 1980 injected a large cloud of ash into the stratosphere but only a small volume of sulfates (0.3 metric tons) and the effect on global climate was negligible. In contrast, the El Chichón eruption of 1982 generated about 20 metric tons of sulfate aerosols from a similar volume of magma to the Mt. St. Helens eruption. Northern-hemisphere temperatures were depressed several tenths of a degree Celsius as a result (Rampino, Self, and Strothers 1988). A far more devastating eruption in 536 A.D., the source of which remains unidentified, caused widespread famine (Anonymous 1992). The explosion of Toba in Sumatra occurred about 73,000 years ago and erupted the equivalent of 2,000–2,800 km³ of magma, far larger than either the Tambora or Krakatau eruptions. A large quantity of H_2S and SO_2 were injected into the atmosphere. This pyroclastic eruption occurred shortly after the earth entered another glacial phase, and Rampino and Self (1992) suggested that the volcanic aerosols may have reduced global temperatures by

THE MOTHER OF MASS EXTINCTIONS

3–5°C, accelerating the onset of glacial conditions. However, once again the global cooling produced no identifiable increase in the number of extinctions.

Sulfate aerosols adsorb and backscatter incoming solar radiation, but the magnitude of associated cooling depends on the volume and chemical composition of the magma erupted (Palais and Sigurdsson 1989; Sigurdsson 1990a, b; Rampino and Self 1982; Rampino, Self, and Strothers 1988). Iron-rich magmas such as basalts may produce massive amounts of sulfate aerosols and release ten times more sulfur than similar-sized silicic eruptions (Palais and Sigurdsson 1989), but the relationship between iron and sulfur content is complex. Furthermore, studies of the 1783 Laki fissure eruption in Iceland indicate that efficient release of SO_2 from basalts requires very vesicular lavas and thus high concentrations of either CO_2 or H_2O (Sigurdsson 1990b). Campbell et al. (1992) have argued that the Siberian flood basalts may have injected substantial amounts of sulfur into the atmosphere, atypically for flood basalts. The mineralogy of the traps, particularly the presence of the world's largest copper-nickel-sulfide ore complex, the presence of extensive anhydrite in the sediments through which the magma ascended, and the large amounts of pyroclastic tuffs associated with the Siberian Traps all suggest that these eruptions were far more violent than most flood basalt flood eruptions, and that they introduced large volumes of sulfate into the atmosphere. These authors suggest global cooling and the development of continental glaciation, induced the end-Permian regression and caused the mass extinction. However, there is considerable evidence that the extinction began prior to the onset of the Siberian volcanism, and little evidence for global cooling. It is not unreasonable that the climatic effects of the volcanism may have been related to the lengthy interval before the recovery began. Sigurdsson (1990b) suggests that injection of 10^{12} kilograms of sulfuric acid would cool the globe by 3–4°C. Just to complicate things further, the climatic effects of volcanic eruptions may not increase linearly with the volume of SO_2 injected. As the volume of aerosol increases, the particles begin to interact and condense into larger particles that actually have less climatic effect (Pinto, Turco, and Toon 1989). Magmas generated above subduction zones may have higher concentrations of sulfur than other magmas and contribute substantial volumes of SO_2 to the atmosphere, as did Mt. Pinatubo in the Philippines. Unfortunately, the anhydrite produced by such eruptions is readily destroyed, making it difficult to recognize high-sulfur eruptions in the fossil record (Luhr 1991).

The climatic effect of a volcanic eruption depends not only on the volume and composition of the material erupted, but also the style of eruption. Global cooling requires the injection of volcanic ash or gases into the stratosphere where the ash or gases circulate around the world. The plinian columns associated with large pyroclastic eruptions make such eruptions very likely to be injected into the stratosphere. In addition, pyroclastic eruptions are several orders of magnitude more voluminous than effusive eruptions. Thus pyroclastic eruptions should have a greater climatic effect than most basaltic eruptions. The conventional wisdom is that flood basalts do not produce plinian columns (which leave rather characteristic deposits) and since they do not inject material into the stratosphere they may have relatively little effect on global climate. However, flood basalts with individual flow volumes greater than 1,000 km^3 and rapid effusion rates, a class that includes the Siberian Traps, may be voluminous enough to inject large quantities of sulfate aerosols into the stratosphere (Strothers et al. 1986).

Clearly the relationships between volcanism and climate are complex and there are few good ways to estimate the volume and composition of gases that may have been produced by past volcanic eruptions, other than by guessing.

Pyroclastic Eruptions

The volcanic origin of the widespread boundary clayrock in South China led Yin Hongfu and his colleagues at the University of Geosciences at Wuhan to emphasize the possible biotic effects of massive volcanism (chapters 2 and 5) (Yin et al. 1989, 1992; Yin, Xu, and Ding 1984; Yang et al. 1987). In addition to the boundary clay itself, there are other ash beds at most localities, suggesting repeated volcanism during the latest Permian. The tuffaceous texture and geochemistry of the boundary clays and the abundant microspherules all support the conclusion that the boundary clays represent marine deposition of an altered intermediate to acidic tuff from silica-rich volcanism.

Less certain, however, is the connection between the volcanism and the mass extinction. Yin Hongfu and his colleagues necessarily argue for a catastrophic rather than a gradual extinction, although at present the preponderance of the evidence is against this pattern. They invoke the formation of a global dust cloud and subsequent elimination of photosynthesis as the cause of the extinction. The abrupt drop in $\delta^{13}C$ in the South China sections is offered as evidence of a catastrophic cessation of primary marine productivity. As noted in chapter 7, the volume of carbon in living organisms is insufficient to cause a whole ocean shift in $\delta^{13}C$ to the extent observed at the Permo-Triassic boundary, and could only cause a loss of the shallow-deep-water isotopic gradient.

Comparative studies of similar events in geologic history cast further doubt on the liklihood that the pyroclastic eruptions are associated with the extinction. Since relatively large pyroclastic eruptions ($> 1,000$ km³) are common in the geologic record (Smith 1960; Christensen 1979), the relationship between the volume of an ash-flow sheet, the size of the plinian column, and the amount of injected aerosols allows a useful test of hypotheses relating pyroclastic volcanism and mass extinctions. In collaboration with Tom Vogel, a volcanologist and igneous petrologist, I tested the biologic effects of pyroclastic volcanism (Erwin and Vogel 1992). We analyzed four extremely large pyroclastic volcanic eruptions with large plinian columns and extensive ash-flow sheets: the Toba Tuff, which 2,800 km³ erupted 75,000 years ago (Chesser et al. 1991); the Huckleberry Ridge Tuff, the largest ash-flow sheet in the Yellowstone Caldera, which erupted 2,500 km³ of material 2.01 million years ago (Christensen 1984); the Tuff of Blacktail and the Blue Creek Tuff in the Heise Volcanic Field in the Eastern Snake River Plain have volumes of 1,500 km³ and 1,400 km³, respectively at 6.5 million years and 6.0 million years (Morgan, Doherty, and Leeman 1984; Morgan and Bonnichsen 1989); and the Elkhorn Mountain volcanics erupted from 75 to 81 million years and the remnants of a volcanic field may have covered 26,000 km² to thicknesses of 4.6 km (Smith 1960; Rutland et al. 1989). Individual ash-flow sheets within the Elkhorn Mountain volcanics appear to be in the range of 1,500–2,000 km³.

Each of these eruptive events is of a similar order of magnitude to the end-Permian pyroclastic eruption but there is no evidence of associated extinctions in the record of global marine diversity, global terrestrial vertebrate diversity, or regional terrestrial diversity. Sepkoski's data on marine familial and generic diversity show an extinction low during the Campanian Stage, when the Elkhorn Mountain volcanics erupted, lows during the early Messinian, and no increase near 2.0 million years (Sepkoski 1989; Raup and Sepkoski 1986a). Likewise, Benton's compilation of terrestrial vertebrate extinction shows no increase during any of the four eruptions (Benton 1987a, 1988).

Such pyroclastic eruptions might cause regional extinctions that are not picked up in global compilations like those of Sepkoski and Benton. Vogel and I also looked at recent analyses of North American terrestrial vertebrate diversity (Stucky 1987). For example, the Black-tail and Blue Creek tuffs erupted during the North American Land Mammal Stage Hemphillian 1 and the Huckleberry Tuff during the Irvingtonian 1, but extinction rates during these stages are not above background rates. A similar study involved the impact of a massive Ordovician eruption of about 1,140 km³ ash covering much of North America and Europe. Again, there was no apparent impact on the marine fauna (Huff, Bergstrom, and Kolata 1992).

This comparison of well-documented pyroclastic eruptions of similar magnitude to the end-Permian event demonstrates that massive pyroclastic volcanism appears to be insufficient to precipitate even regional mass extinctions.

Flood Basalts

In addition to the cooling effects of sulfates, over a longer period the CO_2 from volcanic eruptions can also act as an insulator and initiate a greenhouse effect. This mantle degassing should cause a drop in $\delta^{13}C$ and a shift in $\delta^{18}O$. In 1985 Dewey McLean proposed that release of CO_2 from the Deccan Traps flood basalts in India was the principal cause of the end-Cretaceous mass extinction. In addition, Larson's superplume hypothesis (discussed in chapter 9) suggests that the warm climates of the Late Cretaceous resulted from the introduction of massive amounts of CO_2 into the atmosphere during the production of extensive mid-Pacific flood basalts. If the eruptive rates estimated by Campbell et al. (1992) are correct, enormous volumes of sulfate aerosols and CO_2 may have been introduced into the atmosphere during the formation of the Siberian Traps.

McLean (1985) suggested that the end-Cretaceous mass extinction resulted from buildup of atmospheric CO_2 through a failure of primary productivity and resulting decline in the amount of CO_2 sequestered in the deep ocean. This caused global warming, a lowering of pH of the upper ocean, and perturbation of the ecosystem. He estimated that Deccan traps contributed 5×10^{17} moles of CO_2 over 0.53 to 1.36 million years, increasing the rate of CO_2 release by 10–25 percent. This estimate is highly dependent on the amount of CO_2 in the magma and the fraction of it released during degassing, yet neither of these values is known with any precision. Using a volume three times that of the Deccan Traps and assuming the other values are constant suggests that the Siberian Traps may have injected 1.5×10^{18} moles of CO_2.

In both modern and Cretaceous oceans the upper 100 m of the ocean is reasonably well buffered against pH changes by calcareous microplankton, but the release of this large a volume of CO_2 would have swamped the system and lowered the pH of the waters, perhaps triggering carbonate dissolution. A reduction in the biomass of the calcareous microplankton could have established a positive feedback loop, further increasing CO_2 buildup in the oceans and the atmosphere. During the Permian, calcareous microplankton did not exist, although calcareous algae were abundant until the onset of the regression. Consequently, the buffering system might not have been as well developed as it was by the Cretaceous. Thus the effects of increased atmospheric CO_2 may have been more severe.

However, the calculations at the end of chapter 7 demonstrate that the volume of CO_2 required to produce a shift in $\delta^{13}C$ from $+1°/oo$ to $-3°/oo$ is many orders of magnitude greater than is reasonable, from which I conclude that at best the Siberian traps were only a contributory source of CO_2 during the end-Permian. Finally, invoking either the pyroclastic eruptions or the Siberian Traps as the cause of the extinction requires a catastrophic or, at best, very rapid extinction, a pattern not supported by the fossil record.

David Raup recently suggested another possible link between the eruption of the Siberian traps and the extinction. He noted that if the traps were erupted as rapidly as the latest estimates suggest (about 1 million years), the heat flow would have been tremendous and this alone may have triggered substantial climatic effects. No estimates have been made of how important a factor this heat anomaly may have been.

Trace Element Poisoning

In 1959 Preston Cloud, one of the century's outstanding paleontologists, proposed that an increase in trace-element concentration in the oceans was responsible for the extinction. However, there is little direct evidence for Cloud's hypothesis. More recently, Holser et al. (1991) pointed out that anoxic marine waters during the marine regression might have caused the mobilization of trace metals, thus explaining the trace-element patterns found during the analysis of the Austrian GK-1 core. They suggest that this may have been a contributory cause of the marine extinction. The difficulty with this hypothesis is the enormous amount of trace element that must be added to the oceans to act as a poison. In addition, this mechanism is inconsistent with the taxon-specific patterns of extinction and the terrestrial extinctions.

A SCENARIO FOR THE END-PERMIAN MASS EXTINCTION

MURDER ON THE ORIENT EXPRESS was one of Agatha Christie's masterpieces, in which the victim, a singularly loathsome man, is found murdered in his compartment on the Orient Express with twelve knife wounds. As Hercule Poirot listens to the stories of each passenger on the coach, he finds they eliminate one another as suspects. Despite the abundant and contradictory clues left by the murderer, Poirot is left with a body and two possible solutions. Either an unknown person entered the train, killed the American gangster and made off into the night (the simple, straightforward answer) or all eleven passengers plus the porter (making a good English jury) stabbed the man once each in retribution for his kidnapping and murder of a baby girl many years before (the complicated—though a bit messy—and correct solution).

The situation facing Poirot is not unlike that facing us. Although some suspects can be eliminated immediately (global cooling, species-area effects, extraterrestrial impact), a variety of simple solutions that present themselves as possibilities (oceanic anoxia, climate change caused by massive volcanism, etc.) also suffer from some important problems. It seems reasonably clear that the explanation of the extinction must involve the regression, and obviously must be consistent with the isotopic evidence. The key may lie in one of those messy networks of causality where tugging on any one string leads inevitably to another and another. To a scientist the latter solution has its own problem, for we are trained to prefer the simple, straightforward solution to a more complicated scenario. Indeed this has been codified as the principle of parsimony and Occam's Razor, which states that one should not multiply hypotheses unnecessarily. This simply follows human nature. It is far easier to blame World War I on the assassination of Archduke Ferdinand than on the complexities of Balkan politics, for example.

In the case of the end-Permian mass extinction a more complicated explanation may be required by the evidence. The close of the Permian was a most unusual interval in earth history simply because of the many geologic events that were concentrated during that brief interval. In my view these events were bound to produce an elevated level of extinction and the important question is identifying the factors that contributed to the extreme marine extinction and somewhat smaller terrestrial extinction at the close of the period.

My own view of the causes of the end-Permian extinction is detailed in figure 9.4. I believe that the extinction cannot be traced to a single cause, but rather a multitude of events occurring together, in particular the increased climatic and ecologic instability associated with the regression and a combination of greenhouse warming and possible oceanic anoxia from increased atmospheric CO_2. One of the most striking aspects of the latest Permian is the diversity of inputs of light carbon, including juvenile volcanic CO_2, oxidation of organic carbon on shallow marine shelves, and release of gas hydrates. The last two sources will increase as the regression progresses. I would

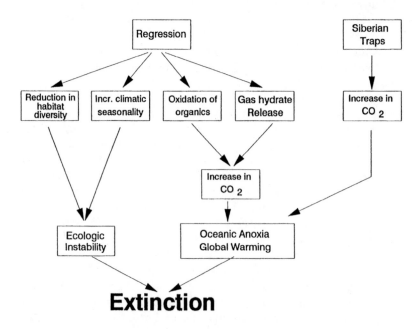

Figure 9.4. The "Murder on the Orient Express Hypothesis," involving multiple causes.

suggest that we need to look at the extinction in three parts: the terrestrial events, the marine events along the margin of Pangea, and the events in South China. The terrestrial extinction was not insignificant, but does not appear to be as extensive as the marine extinctions. Here global warming, increased seasonality, and a possible reduction in habitat diversity associated with harsh environmental conditions in the interior of Pangea seem sufficient to cause the extinction. The major phase of the marine extinction (figure 9.4) appears to have begun earlier on the mainland of Pangea with the onset of the regression and the corresponding loss in habitat diversity and reduction of shelf area. The major extinctions were triggered by the dramatic increase in CO_2 and methane from oxidation of organic material, gas hydrates, and possibly other sources. This triggered global warming and perhaps an increase in oceanic anoxia. By this point the effects were sufficiently far-reaching to affect various refugia, including South China, eliminating many of them. Others persisted and sheltered the Lazarus taxa that were to repopulate the shallow shelves following the amelioration of these conditions during the Early Triassic.

AN AGENDA FOR FUTURE RESEARCH

CLEARLY THIS discussion of the end-Permian extinction has raised as many questions as it has answered and I have identified areas requiring further study in the chapters on the boundary (chapter 2) and the extinction patterns (chapters 3 and 4). Here my concern is with the work required to achieve a better understanding of the mechanisms of extinction. In part this depends on the improvement of intercontinental biostratigraphic correlations and higher-quality work analyzing patterns of extinction, as discussed previously. Many of the studies cited on previous pages have been conducted on a very coarse stratigraphic scale. The fine-scale work on the Austrian GK-1 core demonstrates what can and must be done to resolve patterns at the appropriate scale. Stable isotope studies have been a boon to the study of mass extinctions, as they have to so many other areas of geology, but we still have few constraints on the mechanisms responsible for the observed shifts. Likewise, volcanologists and mantle geochemists have provided few constraints on speculations concerning the possible involvement of mantle plumes with the extinction. Are all mantle plumes geochemically and isotopically similar? Are there new techniques that may allow reasonable estimates of the amount of CO_2 and SO_2 released by flood basalts. The Siberian Traps do not seem to be associated with a mantle plume. Does this suggest a shallow mantle source for this flood basalt (and thus perhaps an impact as a trigger), or is this fortuitous?

These are but a few of the questions that need answering, certainly many others may have occurred to the reader. This plethora of questions demonstrates how much we still have to learn about this episode, and that the answers may lie not in speculation but in gathering new data and seeking new methods of analysis. To bring this discussion to a close we need to move from the extinction itself to the aftermath and to the long-term evolutionary effects of the extinction.

X

Aftermath and Implications

This is the world's limit that we have come to;
this is the Scythian country, an untrodden desolation.
— AESCHYLUS, *PROMETHEUS BOUND*

THE DESIGNATION of the earliest Triassic as the Scythian was most appropriate. Like the Scythian country, the interval following the extinction was also a time of desolation. But the Scythian also saw the beginnings of the recovery and the diversification of many new groups. For example, *Miocidaris* is one of two echinoid genera to have survived the end-Permian mass extinction. This highly opportunistic, widely distributed echinoid (properties that doubtless contributed to its survival) was one of only six genera known from the Permian, so the group might well have suffered the same fate as other low-diversity groups like the trilobites and disappeared. Yet *Miocidaris* survived the extinction and its descendants went on to become one of the most important members of marine communities during the subsequent rebound. There is another, more curious aspect to the survival of *Miocidaris*, however. The test, or shell, of echinoids is composed of two sets of plates, the five files of ambulacrals that cover the tube feet (homologous to the arms on a starfish) and the interambulacrals between them. The number of columns of interambulacral plates is highly variable in most Paleozoic echinoids, but in *Miocidaris* the number is reduced to two columns of plates within each ambulacrum and two columns within each interambulacrum. Since all post-Paleozoic echinoids are descended from *Miocidaris*, the survival of the genus fixed the relationship between ambulacral and interambulacral plates in echinoids (Paul 1988). Was the survival of *Miocidaris* fortuitous? Or did this morphologic innovation aid in the survival of the genus? Whichever is the case, this radically altered the morphology of all later echinoids, limiting future evolutionary pathways and the subsequent course of echinoid evolution (Jablonski 1989). Jablonski (1989:117) summed up this position well:

> Mass extinctions can break the hegemony of species-rich, well-adapted clades and thereby permit radiation of taxa that had previously been minor faunal elements: no net increase in the adaptation of the biota need ensue. Although some large-scale evolutionary trends transcend mass extinctions, post-extinction evolutionary pathways are often channelled in directions not predictable from evolutionary pathways during background times.

The story of *Miocidaris* is one of decimation and recovery, recovery taking the clade in a new evolutionary direction. This is the major theme of this chapter: the postextinction recovery and the vast suite of evolutionary opportunities created by the removal of many once-dominant taxa and the expansion of other clades.

Earliest Triassic ecosystems were more vacant than at any time since the Cambrian metazoan radiation. The Cambrian experienced an explosion of durably skeletonized phyla and a plethora of novel organisms that rank as phyla. Yet no new phyla or classes appear in the Triassic (Erwin, Valentine, and Sepkoski 1987). In fact, the pattern of appearance of new basic morphologies is highly asymmetrical: morphologies accorded phylum, class, and ordinal rank are overwhelmingly clustered in the lower Paleozoic. This asymmetrical pattern of origination is highly significant for it suggests that despite dramatic decline in the occupation density of most adaptive zones during the end-Permian mass extinction, the pattern of extinction was insufficient to permit major morphologic innovations. In part this is doubtless due to the fact that no adaptive zone was entirely vacant. Hence the presence of these persisting species limited the success of broad evolutionary jumps. This comparative study raises an alternative possibility, however. Major morphologic innovations may become less likely over time because of intrinsic changes in developmental and regulatory systems—thus innovations like those of the Cambrian may no longer be possible (Erwin forthcoming).

Recall from the prologue that we left John Phillips trapped in a dilemma of his own making. While the striking contrast between fossils of the Paleozoic, Mesozoic, and Cenozoic formed the basis for Phillips's recognition of the Phanerozoic eras, he also acknowledged faunal continuity at higher levels, particularly in the progressive Paleozoic expansion of groups that became the dominant members of Mesozoic oceans. Which of these views is more accurate? Is the expansion of the Modern Fauna largely a consequence of differential survival during the end-Permian extinction and subsequent expansion, or are the clades of the Paleozoic Evolutionary Fauna united by a higher rate of evolutionary turnover that compelled their eventual decline? The view that the Modern Fauna was destined for success even without intercession from mass extinctions, while admittedly counterintuitive, finds considerable support from studies of the geographic and temporal dynamics of Paleozoic communities and from simulation studies. Although much of the story of the Mesozoic rebound lies beyond the scope of this book, an overview is a necessary prelude before we turn to the broader context of the significance of the extinction in the history of life, particularly the issues raised by John Phillips in 1860.

AFTERMATH: THE MESOZOIC RADIATION

THE END-PERMIAN mass extinction triggered an explosion in marine diversity and a reorganization in marine communities. This reorganization mushroomed through the Mesozoic as a variety of new groups of animals appeared and the breakup of Pangea expanded the number of marine provinces (Valentine 1973a). This established the clades of the Modern Evolutionary Fauna as the dominant elements in most marine communities. This reorganization is so sweeping that Vermeij (1977, 1987) described it as the "Mesozoic Marine Revolution." Similar changes occurred on land during a tremendous burst of evolution among the therapsids in the Triassic, followed by the appearance of dinosaurs late in the Triassic.

The recovery after the end-Permian mass extinction can be divided into three phases of highly uneven length (figure 10.1). The first was a low-diversity interval during the Scythian. Early Triassic sequences in Tethys, the western United States, and elsewhere are characterized by abundant stromatolites, a characteristic "disaster" form (Schubert and Bottjer 1992). Stromatolites are layered sedimentary structures formed by bacteria and cyanobacteria. They were common pre-Cambrian fossils but declined in abundance during the late Proterozoic (about 900 million years ago) as animals began to diversify, presumably because stromatolites are a wonderful food source for grazers. Today stromatolites are confined to harsh environments where animals are excluded, like high-salinity lagoons. The presence of stromatolites at the beginning of the lag phase following the extinction is indicative of either extreme environmental conditions or low abundances of animals (or both). The low-diversity acritarch assemblages associated with earliest Triassic palynofloras (chapter 4) are also typical of the aftermath of an environmental crisis.

This burst of stromatolites lasted only a short time. Later Scythian oceans were populated by large numbers of a few species, with ammonites, bivalves, and brachiopods the most abundant groups. Many of these genera are distributed almost worldwide, in contrast to the high faunal provinciality of the Late Permian; only 3 to perhaps 5 marine

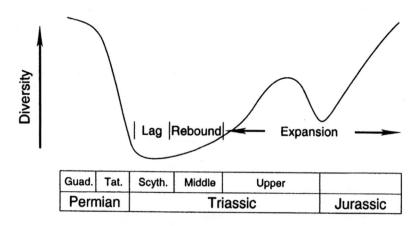

Figure 10.1. Division of the postextinction recovery phase into an immediate lag phase, a rebound phase beginning in the late Scythian and characterized by return of the Lazarus taxa, and an expansion phase with diversification of a variety of new groups.

faunal provinces can be identified. Reefs are almost wholly absent. Kummel (1973a, b) recognizes 136 genera of Scythian ammonites, but only 7 of nautiloids. Bivalves are widespread but of low diversity (see also Yin 1985a).

> The other invertebrate groups such as foraminifers, sponges, bryozoans, gastropods, crinoids, echinoids, asteroids, and ophiuroids are extremely rare in the fossil record of Scythian rocks. It thus can readily be seen that Scythian marine faunas are unusual in their lack of diversity and representation. The extreme rarity and fragmentary nature of the record for many of these invertebrate groups inhibits clear understanding of the evolutionary relationships to their Permian ancestors. (Kummel 1973b:226)

Early Triassic communities exhibit low diversity and a very simple structure in comparison to Permian or later Triassic communities, a feature often associated with communities enduring environmentally demanding conditions, although it might simply reflect the aftermath of the extinction. On land, the *Lystrosaurus* vertebrate assemblage is distributed worldwide and is likewise comprised of only a few species.

These depauperate faunas are hardly surprising, although the lag between the end of the extinction and the onset of the recovery phase lasted proportionally longer than lags following other mass extinctions. (Calculating the duration of the lag is hampered by uncertainties over the age of the boundary. The duration of the Early Triassic is traditionally considered to be 4 million years [Harland et al. 1989], but new dates for the base of the Triassic may add up to 5 million years to the interval, see chapter 3.) This lag suggests a continuation of harsh environmental conditions through the Early Triassic. Alternatively, if the rebound followed a traditional logarithmic curve the Scythian may represent a biologically controlled lag prior to the exponential phase of the radiation (e.g., see Sheehan 1985, on reefs), although if this were the case we would not expect virtually simultaneous rediversification across many clades. As Stanley (1990) notes, the lack of carbonate deposition and reefs suggests inclement environmental conditions were primarily responsible for this lag.

The second phase of the recovery began with the return of normal marine faunas during the Middle Triassic, but these faunas are an interesting combination of so-called Lazarus taxa returning from the as yet unrecognized refugia in which they survived the Late Permian and Early Triassic and new, Triassic groups. Gastropods are one of the best examples of this phenomenon. In 1973 Roger Batten noted 32 genera and 16 families of Guadalupian affinities are found in the Ladinian but have never been discovered in the Dzulfian or Scythian. These Paleozoic holdovers begin to diversify before new Triassic lineages (Erwin 1990a). Accordingly, Ladinian gastropod assemblages are more similar to those of the Guadalupian than of the Jurassic. The combination of the Carnian-Norian radiation and the end-Triassic extinction finally eliminated Paleozoic gastropods and imparted a modern look to the gastropod faunas (Erwin 1990a). The many Lazarus taxa lend further credence to the argument that harsh environmental conditions inhibited survivors of the extinction from radiating until conditions improved.

The evolutionary history of reefs is an excellent example of the complex dynamics during the early phases of the postextinction recovery. Reef ecosystems are particularly sensitive to environmental instabilities and are consistently decimated by mass-extinction events (Sheehan 1985). As we saw in chapter 7, the end-Permian mass extinction was no exception to this rule. The calcisponge *Girtyocoelia*, a common member of Permian reefs, is common in Early Triassic rocks (Flügel and Stanley 1984) and is only one of several such reef-building organisms identified. When reefs reappear in the Middle Triassic they are populated by a combination of Paleozoic holdovers, including calcareous algae and calcisponges, together with a number of new groups. The first reefs of the Anisian are small-scale carbonate mounds that are quite similar to Permian reefs in both biotic content and structure (G. D. Stanley 1988; see also S. Stanley 1988b) and most Middle Triassic reefs are dominated by holdovers from the Paleozoic. Among the most important of the new groups are the scleractinian corals, a group only distantly related to the tabulate and rugose corals of the Paleozoic (Flügel and Stanley 1984), but since the return of reefs did not require the evolution of new reef specialists like the scleractinians, it appears that the lag was not due to biological factors, but rather inhibition by climatic factors or ocean chemistry.

The third and longest phase of the recovery began in the Carnian as diversification accelerated, progressively displacing groups of Paleozoic affinities with new groups. This expansion phase was interrupted by the end-Triassic extinction, but resumed in the Jurassic and continued into the Cretaceous with the appearance and diversification of neogastropods, several arthropod clades, and many other new groups.

FROM THE PALEOZOIC TO THE MESOZOIC: SUDDEN CHANGE OR GRADUAL TRANSFORMATION?

THE DEBATE over the transformation from the faunas of the Paleozoic to those of the Mesozoic, and indeed largely those of the entire post-Paleozoic in the marine realm, is at the heart of a much larger controversy over the controls on the diversification of life. Paleobiologists concerned with this issue have focused on a number of questions, chief among them: Is the apparent pattern of increasing diversity (figure 1.3) real, or is it an artifact of an increasingly good record toward the present? Assuming the pattern is real, is it driven by physical patterns, such as increased provinciality (Valentine and Moores 1972; Valentine 1980) or continuing physical change (Cracraft 1985), by biotic factors such as a reduction in the probability of extinction (Raup and Sepkoski 1982; Flessa and Jablonski 1985), an increase in adaptive space through time (Bambach 1983, 1985), or progressive improvement within lineages (Vermeij 1987; Sepkoski 1984), or is the pattern devoid of meaning in the grand sense, simply reflecting events at smaller scales (Hoffman 1986, 1989a; Hoffman and Ghiold 1985)? (See Benton 1987a, c, 1990a; Signor 1990 for recent reviews of these issues.) It is now apparent that the pattern of increasing diversity through time is real, at least in broad outline (Benton 1990a; Signor 1990), but there is little consensus among paleontologists on which of the possibilities listed above are responsible for the trend. The debates over these questions focus on long-term evolutionary trends through the Phanerozoic that lie beyond the purview of this volume, but the effect of the end-Permian mass extinction is central to many of the discussions.

Sepkoski's description of the three Evolutionary Faunas (chapter 1) was part of his development of equilibrium diversity models to explain these long-term diversity trends (Sepkoski 1979, 1981, 1984). These models are derived from logistic models of equilibrium theory in ecology, and include MacArthur and Wilson's theory of island biogeography discussed in the preceding chapter. In this model each evolutionary fauna has a characteristic rate of origination and extinction. Early in their history each fauna diversified exponentially. In a manner analogous to the theory of island biogeography, each fauna also has an equilibrium family diversity where the rate of familial origination equals the rate of familial extinction. The exponential diversification of each fauna slowed as the equilibrium diversity of the fauna was approached. The replacement of the Cambrian Evolutionary Fauna by the Paleozoic Evolutionary Fauna is explained by a slower initial rate of growth coupled with a higher equilibrium diversity. The higher equilibrium diversity reflects lower intrinsic rates of origination and extinction among the Paleozoic Evolutionary Fauna. Similarly, the replacement of the Paleozoic Evolutionary Fauna by the Mesozoic Fauna was, in this model, an inevitable consequence of still lower initial rates of diversification and a higher equilibrium diversity. The Modern Fauna continues to expand and shows no sign of reaching an equilibrium point.

In this model the eventual success of the classes that comprise the Modern Fauna was determined long before the end-Permian mass extinction. Sepkoski (1984) notes that the Paleozoic Fauna maintained a high diversity into the Middle Devonian, whereupon a slow decline in familial diversity began until the Paleozoic fauna was decimated during the end-Permian mass extinction. In contrast, the Modern Fauna continued to expand throughout the upper Paleozoic (compare figures 1.5 and 1.6). As pointed out in chapter 3, the Paleozoic fauna suffered a 79 percent extinction at the close of the Permian while the Modern Fauna declined only 27 percent (Sepkoski 1984). Sepkoski views this differential impact during the mass extinction as part of a general pattern that older evolutionary faunas have higher rates of extinction than younger faunas during mass extinctions. Sepkoski modeled the diversification history of the three Evolutionary Faunas with a three-phase kinetic model incorporating differences in intrinsic rate of growth. After incorporating time-specific perturbations into the model as proxies for mass extinction events he was able to simulate the changes in faunal dominance as a consequence of the extinctions. Sepkoski emphasizes:

> The perturbations in the model merely accelerate the changes in phase composition [from one evolutionary fauna to the next] and do not produce permanent changes that would not have occurred in the absence of perturbations. . . . Even without the great Permian mass extinction, the change in dominance would have occurred during the Mesozoic-Cenozoic interval and the class level composition of the fauna in the present oceans would have appeared very much the same. (Sepkoski 1984:262).

Empirical studies by Sepkoski and his colleagues have lent support to this model. For example, Sepkoski and Miller (1985) compiled data on the environmental distribution of over 300 North American Paleozoic communities. They determined the position of each assemblage along a simple, two-dimensional, onshore-offshore gradient, and analyzed patterns of generic diversity within taxonomic orders. Their results suggest that the Cambrian, Paleozoic and Modern Evolutionary Faunas each occupied distinct environmental areas during the Paleozoic and, moreover, that the Modern Fauna was steadily displacing the Paleozoic Fauna during the Upper Paleozoic (figures 10.2 and 10.3). In detail, their results suggest a pattern of onshore origination for each of the three faunas and progressive displacement of older faunas to more offshore environments (see also Miller 1988, 1989; Jablonski and Bottjer 1990).

This equilibrium diversity approach has been subject to criticism from a variety of quarters. Hoffman (1985a) and Cracraft (1985) rejected as unsound the necessary extrapolation from the ecological level at which the theory of island biogeography was developed to the global evolutionary scale. Cracraft (1985) argued that diversification was controlled by diversity-independent factors, not the diversity-dependent factors invoked by global equilibrium models.

A nonequilibrium approach was championed by Kitchell and Carr (1985). As with Sepkoski's approach, Kitchell and Carr's model is diversity-dependent, but differs in that no equilibrium diversity level is achieved. They find that continuing perturbations by mass extinctions and evolutionary innovations may have maintained global diversity in a continuing nonequilibrium state. Despite the differences in the structure of their model, Kitchell and Carr also conclude that faunal turnover between the Paleozoic Evolutionary Fauna and the Modern Evolutionary Fauna would have been only slightly delayed (in geological time) if the end-Permian mass extinction had not occurred. The diversity-dependent assumptions of both equilibrium and nonequilibrium models have been criticized (Walker 1985; Cracraft 1985; Hoffman 1985a) as assuming a fixed number of ecological niches that are generally filled. In the absence of a mass extinction, invading species can succeed only by displacing a species already occupying a particular niche.

One of the major difficulties with both equilibrium and nonequilibrium diversity models is that they remain essentially descriptions of pattern without any discussion of evolutionary process. As such they are not necessarily incorrect, but they are inherently less satis-

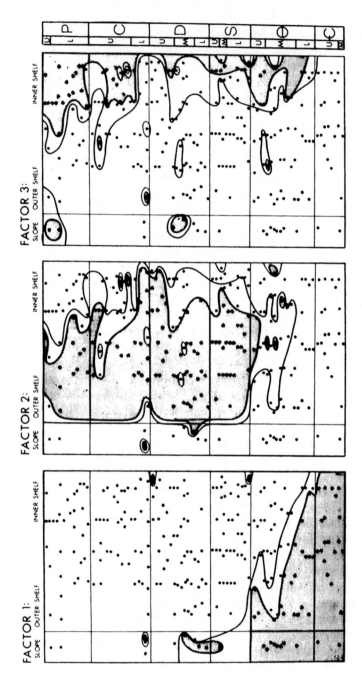

Figure 10.2. Results of Sepkoski and Miller's (1985) Q-mode factor analysis of Paleozoic community diversity patterns along a simplified, two-dimensional onshore-offshore gradient. These are contoured time environment diagrams; time is along the vertical axis and the onshore-offshore gradient along the horizontal axis. The dots represent individual communities. The shaded portion of each plot shows the area of greatest concentration of communities dominated by the Cambrian Evolutionary Fauna (Factor 1), the Paleozoic Evolutionary Fauna (Factor 2), and the Modern Evolutionary Fauna (Factor 3). The contours enclose the 0.33 and 0.67 divisions. From Sepkoski and Miller (1985).

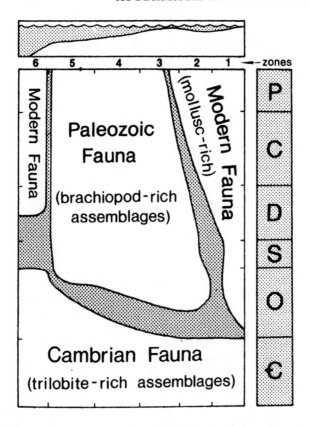

Figure 10.3. An interpreted summary diagram of figure 10.2, showing the successive expansion of each new Evolutionary Fauna from a nearshore position and displacement of older faunas offshore. From Sepkoski (1991).

fying than a testable model of the evolutionary processes that produce the patterns. As Benton (1990a) emphasizes, the available data are insufficient to choose between the equilibrium and nonequilibrium diversity models. Both models share a view of the end-Permian mass extinction as essentially irrelevant to the history of life, a view seemingly confirmed by differences in turnover rates between different Evolutionary Faunas (and, by extension, between their constituent clades) and onshore-offshore diversification patterns.

What of the alternative position, that the end-Permian mass extinction was the pivotal event in the elimination of those groups that dominated Paleozoic marine communities and allowed, perhaps even fostered, the radiation of the new clades that came to comprise the Modern Evolutionary Fauna?

In the discussion of extinction patterns in chapter 3, I emphasized a number of taxa, including the asteroids, echinoids, bryozoans, and articulate brachiopods for which the end-Permian mass extinction appears to have been central to this transformation.

Certainly for the group I know best, Paleozoic gastropods, the extinction was critical to their history (Erwin 1990a). Recall that the Class Gastropoda is a part of the Modern Evolutionary Fauna, based on the tremendous diversification of the group since the Cretaceous. Yet in many ways the history of the group forms a microcosm of Phanerozoic diversity. Paleozoic gastropods include a variety of groups quite unlike modern gastropods in morphology and far fewer were predators and far were more herbivores, filter-feeders, and detritus-feeders. Although the group as a whole suffered comparatively little during the extinction, such was not the case for individual groups. For example, the sessile filter-feeding gastropods, including members of the Family Omphalotrochidae, were wiped out by the extinction as were the platyceratids, a group that was basically ectoparasites on crinoids. A very new group of gastropods diversified in Middle and Upper Triassic, one that is far more modern in aspect than the Upper Paleozoic gastropods. Moreover, the ecological variety of the group makes them a wonderful source of insight into patterns of ecological selectivity during the extinction. This has allowed me to use the taxonomic history of the group to determine the importance of the extinction in the elimination of the Paleozoic groups and the diversification of the Triassic gastropods. I performed a factor analysis of generic diversity from the Carboniferous through the Triassic. My results (figure 10.4) demonstrate that the extinction was the pivotal event in the transition between the Paleozoic and Mesozoic. There is no indication that the Paleozoic forms were already on their way out before the extinction, nor increasing diversity among those groups that later became important members of the Triassic gastropod assemblage. From this I concluded that, for gastropods at least, there is little support for the contention that this event was largely irrelevant to long-term trends in the history of life.

Van Valen (1984) noted that the probability of extinction declines exponentially on a plot of the probability of family-level extinction per million years versus time through the Phanerozoic. Rather than

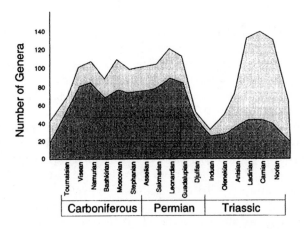

Figure 10.4. Results of a Q-mode factor analysis of gastropod generic diversity from the Carboniferous through the Triassic. The clades belonging to factor 2 (light stipple) have relatively constant diversity through the Upper Paleozoic and suffer more heavily than the members of the Paleozoic assemblage (black) during the end-Permian mass extinction. Despite this they account for the bulk of the Mid- to Upper Triassic radiation. From Erwin (1990a).

declining continuously, however, as would be expected from Darwin's argument that the absolute fitness of taxa should be steadily increasing, an abrupt shift occurs at the Permo-Triassic boundary. After the extinction the probability of extinction returned to the same level as the Ordovician before beginning a second decline. Van Valen interprets this as a resetting of the clock of community evolution as a consequence of the mass extinction. This issue, like so many others remains open.

The discovery by Jack Sepkoski and David Raup of apparent periodicity in the record of mass extinctions over the past 250 million years adds further complications, for the periodicity hypothesis suggests that mass extinction events may provide more structure to the history of life than the ongoing process of adaptation. This possibility remains a highly contentious issue, but since the end-Permian mass extinction is the first event in the periodic cycle detected by Raup and Sepkoski, the hypothesis will be discussed briefly, particularly the similarities and differences between the end-Permian and end-Cretaceous mass extinctions.

271

THE PERIODICITY HYPOTHESIS

CATASTROPHISM REENTERED the lexicon of reputable geologists in 1980 with the publication of the Alvarez impact hypothesis (see preceding chapter). Four years later David Raup and Jack Sepkoski suggested that extraterrestrial impacts might be a regular feature of the history of the earth when they presented evidence for periodic mass extinctions approximately every 26 million years over the past 270 million years, beginning with the end-Permian mass extinction (Raup and Sepkoski 1984). Their work expanded on earlier suggestions (Fisher and Arthur 1977) of a 32-million-year cycle in extinction rates, but Raup and Sepkoski's data were derived from Sepkoski's compendium of Phanerozoic marine family diversity. The initial analyses showed 12 peaks (varying considerably in magnitude), including the end-Permian, end-Triassic, and end-Cretaceous mass extinctions. Later analysis of Sepkoski's generic database (Raup and Sepkoski 1984; Sepkoski and Raup 1986; Sepkoski 1986a, b) confirmed a periodic pattern with 9 peaks and a period length of 26.2 million years (figure 10.5) (Sepkoski 1989). The periodicity model has been heavily criticized by systematists (Patterson and Smith 1989; Hoffman 1985a, 1989c, d; Hoffman and Ghiold 1985), statisticians (e.g., Quinn 1987; Stigler and Wagner 1987) and others (e.g., Baksi 1990), with various rejoinders from Raup and Sepkoski (see particularly Sepkoski 1989, 1990). (In an amusing sidelight, during a dispute that at times became unnecessarily vituperative, Sepkoski [1986a] demonstrated that one of Hoffman's [1985b] criticisms actually strengthens the periodic signal, rather than weakens it, a fact Hoffman had not realized.) As I write this chapter in 1992, the controversy has settled down a bit and the periodic signal continues to shine through the turmoil, battered but resilient.

Raup and Sepkoski had difficulty identifying a plausible terrestrial cycle with a 26-million-year period. The inclusion of the end-Cretaceous extinction in the series led to an immediate association between periodic mass extinctions and the impact hypothesis. Reports of a similar pattern in terrestrial impact structures (Rampino and Strothers 1988) lent support to the association. As we have seen, there is little support for an extraterrestrial impact at the Permo-Triassic boundary. There is well-known evidence for impact events at the end of the Cretaceous and at the Eocene-Oligocene boundary. More recently, increased iridium levels have been found at the end-

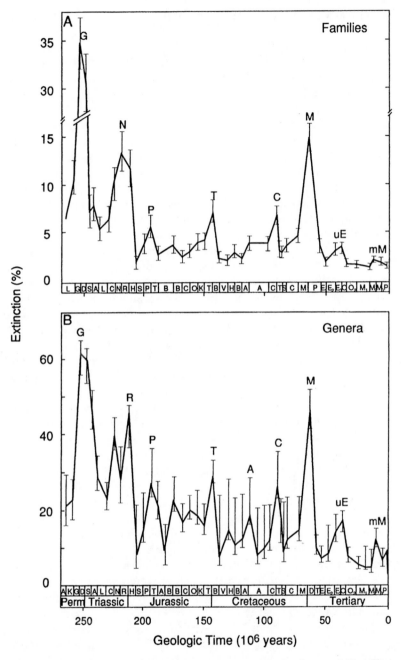

Figure 10.5. Familial and generic extinction patterns for the past 250 million years, showing an apparent periodicity of extinction peaks every 26 million years. From Raup and Sepkoski (1986a). Copyright 1986 by the American Association for the Advancement of Science.

Ordovician boundary in China but these seem to be related to a decrease in the sedimentation rate rather than an extraterrestrial impact (Wang et al. 1992). Microtektites have been found associated with the Late Devonian mass extinction (Wang 1992) and shocked quartz has been found associated with latest Triassic faunas in Italy (Bice et al. 1992). If the periodic signal is real, then the forcing agent must be terrestrial, involving mantle processes and expressed as mantle plumes (Loper, McCartney, and Buzyana 1988; Loper and McCartney 1990; McCartney, Huffman, and Tredoux 1990) or magnetic reversals (Raup 1985; Strothers and Rampino 1990), and the association with impacts is fortuitous. Alternatively, the forcing agent may be extraterrestrial but evidence of impact has not be discovered at other boundaries.

One difficulty with the periodicity hypothesis is the pronounced differences between the end-Permian and end-Cretaceous mass extinction, the two largest events in the periodic series. If the periodicity hypothesis is substantially correct, we should expect some similarities between the two events. Comparison of these events is hampered by two factors. First, the extinctions occurred under different conditions and the same mechanism would be unlikely to produce closely similar extinction patterns. Nonetheless, strong dissimilarities certainly should sound a cautionary note for advocates of the periodicity hypothesis. Second, *if* the cause of the periodicity is terrestrial, the end-Cretaceous mass extinction signal may have been sufficiently modified by the bolide impact to obscure the similarities with the end-Permian extinction.

The differences between these two major extinction events can always be explained away by the argument that the differences in geological setting (the position of the continents), climate, biotic diversity, and other factors makes it unlikely that similar extinction patterns would be produced even under virtually identical mechanisms. True to a point, but the differences between the two events and the lack of any strong evidence supporting a connection between them suggests caution. Clearly the strength of the periodicity hypothesis warrants continued investigation (pro and con) but there seems little reason at present to claim that both events were produced by the same causal mechanism.

The debate over the periodicity hypothesis continues to produce a voluminous literature, both pro and con. I do not intend to review this literature here, but the periodicity hypothesis does raise two issues of concern to the end-Permian mass extinction. First, it sug-

AFTERMATH AND IMPLICATIONS

gests that some common forcing agent lies behind each event in the series, and, a bit less forcefully, that the cause of one event may shed light on other events (Sepkoski 1990). Second, the hypothesis adds another alternative to the progressive change-catastrophist dichotomy discussed above. If mass extinctions occur periodically and if they fundamentally reorganize the earth's biota in ways that would not occur absent the extinctions, they may have a profound effect on the history of life and our view of the evolutionary process. As Stephen Jay Gould (1985:8) said: "Mass extinctions are more frequent, more rapid, more extensive in impact, and more qualitatively different in effect than our uniformitarian hopes had previously permitted most of us to contemplate. . . . If anything like progress accumulates during normal times . . . the vector of advance may be derailed often enough and profoundly enough to undo any long-term directionality."

Evidence in favor of this position was advanced by Jablonski's (1986a) analysis of survival patterns of bivalves and gastropods in the Gulf Coast during the Late Cretaceous and across the Cretaceous-Tertiary boundary. He identified a set of traits that increase survival during the intervals between mass extinctions, including planktotrophic larval development, species-richness within genera, and broad geographic range of a species. In fact, there is what Jablonski calls a synergistic effect between species richness and geographic range during background times, in which the combination of the two confers greater survival potential than either alone. Under the "field of bullets" scenario none of these features would enhance survival, while under the "fair game" scenario mollusc clades that possessed these features would have a greater likelihood of survival. In fact, none of these features enhanced survival during the mass extinction. Rather, broad geographic range at the generic level enhanced the probability of survival of the clade. Neither geographic range of the species nor species richness of the genus had any effect. The kicker is that geographic range of the genus is not subject to selection: a genus may have many narrowly distributed species or a few broadly distributed ones. In either case broad geographic range of the genus results, enhancing survival probability. Subsequently, Jablonski has extended these results to other regions and other clades (Jablonski 1989) for the end-Cretaceous extinction.

The critical question is the generality of Jablonski's findings: is the "wanton destruction" scenario a general feature of mass extinctions? If so, this implies a far less important role for adaptive evolution in

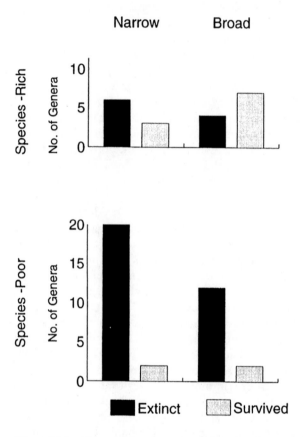

Figure 10.6. Comparison of species richness and geographic range of gastropod richness during the end-Permian mass extinction. From Erwin (1989b).

structuring the history of life. Alternatively, this pattern may be unique to the end-Cretaceous mass extinction. This question remains an area of active research, but when I examined survival patterns among Permian gastropods I found no support for Jablonski's hypothesis (Erwin 1989b, 1990a). In a paleoecological analysis of gastropod assemblages in the southwestern United States, I discovered that during the middle Permian (Leonardian and lower Guadalupian) those genera that were to survive the extinction were distributed across more environments, and were more species-rich than genera that became extinct. When I assessed the geographic range of the two groups by comparing the southwestern U.S. assemblage with an assemblage of the same age from Malaysia, those genera found in both regions had a higher survival rate than genera found only in one. More importantly, the synergistic effect of species richness and broad geographic range that Jablonski found to confer increased survival rates during background times continued to increase survival during the end-Permian mass extinction (figure 10.6), unlike the situation for the end-Cretaceous mass extinction. This analysis suggests that survival of gastropod genera during the mass extinction was enhanced by the same characteristics that enhanced survival during background intervals: species richness, broad environmental distribution, and broad geographic distribution.

The scenario sketched by Jablonski raises the intriguing (at least to paleobiologists) or infuriating (to other biologists) suggestion that survival during a mass extinction is determined by factors beyond the purview of natural selection. If survival during a mass extinction is generally controlled by the geographic range of a genus (however acquired), then adaptation during the intervals between mass extinctions is of far less significance than previously thought, for natural selection cannot "see" the geographic range of a genus. Until recently extinction has been viewed as a negative process, culling the unfit and perhaps advancing evolutionary trends. But this new view of mass extinction suggests a far more positive role in structuring the history of life. Extinctions may fundamentally reorder the relationships among different groups by preserving some and eliminating others on the basis of traits that have little adaptive value to the species.

So what is the answer to the question posed by Phillips and his colleagues so many years ago? The simple fact is that we do not yet understand the importance of the end-Permian mass extinction for the history of life. It is clear that the biggest question of all, what difference did it make, remains open. The dilemma that confounded John Phillips remains unresolved today.

Abundance: The number of different organisms of an individual species.

Adaptation: A feature or attribute of an organism with a useful function.

Adaptive radiation: The rapid production of new species within a clade following invasion of a new geographic region, or exploitation of a new ecologic opportunity.

Albedo: The amount of solar radiation reflected from the earth's surface. The albedo is higher for ice than land, and higher for land than water.

Anoxia: Lack of oxygen.

Arthropod: A phylum of invertebrate animals with jointed legs and a hard skeleton, including crabs, insects, etc.

Basalt: A fine-grained, mafic extrusive rock; the primary component of oceanic crust.

Benthic: Living on the sea floor.

Biogeography: The study of the geographic distribution of plants and animals.

Biomass: The total mass or living material in a particular area or at a particular time.

GLOSSARY

Biostratigraphy: The use of fossils to correlate between rocks in different areas, and for arranging sedimentary units in relative temporal order.

Biota: All the plants and animals with a region.

Bivalve: A clam, and a member of the molluscan class Bivalvia.

Bolide: An extraterrestrial object (e.g., meteorite, comet) that hits the earth.

Boundary Section: A geologic outcrop spanning a geologic boundary.

Brachiopod: A phylum of invertebrate animals with a shell, generally calcareous, and a lophophore for a feeding structure.

Bryozoan: A phylum of colonial marine invertebrates that also have a lophophore, like brachiopods, but evert the lophophore to feed.

Carbonate: Minerals, rocks or, sediments composed of one carbon atom and three oxygen atoms ($CaCO_3$), including the minerals calcite, aragonite, and dolomite.

Carnivore: An animal that feeds upon other animals.

Cenozoic: The most recent era of geologic time, extending from the close of the Mesozoic about 65 million years ago to the present.

Cephalopod: A class of molluscs including squids, octopus, and the extinct ammonoids.

Clade: A group of species all descended from a common ancestor.

Class: A level of the taxonomic hierarchy below phylum and above order.

Clay: Sedimentary particles smaller than 1/256 mm. Also, a member of the clay mineral family.

Community: An ecologic association of populations sharing a common habitat.

Competition: An ecologic interaction between two different organisms both attempting to exploit a common, limiting resource, such as space or food.

Conodont: An unusual group of fish-like organisms, closely related to vertebrates, that produced phosphatic tooth-like structures of great use in biostratigraphy.

Continental drift: The movement of continents relative to one another via plate tectonics.

Core: The center part of the earth below 2,900 kilometers; largely iron and nickel, with a solid inner core, and a molten outer core.

Correlation: The use of fossil, lithologic, magnetic, chemical, or other evidence to determine that two stratigraphic units are of similar age.

Deposit feeder: An animal that feeds by eating sediment and digesting the organic material within the sediment.

Diversity: The number of different kinds (species, genera, etc.) of organisms.

Diversity-dependent extinction: Extinction mechanisms that act to reduce diversity through the depletion of required ecological factors, such as oxygen or habitat area.

Diversity-independent extinction: Extinction mechanisms that reduce diversity of all species equally, without respect to numbers of individuals.

Durably skeletonized: With a skeleton sufficiently robust to be frequently preserved in the fossil record.

Echinoderm: A member of the Phylum Echinodermata, including starfish, sea urchins, and crinoids.

Epicontinental sea: A shallow continental sea.

Epifaunal: Organisms that live attached to, but above, the substrate.

Energy-balance model (EBM): The simplest class of climate simulations, focusing on the balance between incoming and outgoing solar radiation. While this class of models focuses largely on temperature and ignores many other climate variables, these simulations are easier to run and do provide interesting insights.

Evaporites: Salts deposited by evaporation of seawater.

Evolutionary fauna: A grouping of marine classes sharing a common diversity history at the family (and presumably) species level.

Extinction: The total disappearance of a species or taxon.

Flood basalt: Vast accumulations of horizontal basalt flows that cover large areas and generally erupt over a short time span.

Foraminifera: Single-celled animals with a calcareous skeleton, or test belonging to the Subclass Sarcodina. Generally microscopic in size, but sometimes larger.

Fusulinid: One group of foraminifera with long, cylindrical skeletons. Common through much of the Permian and very useful in biostratigraphic correlation.

Gastropod: Scientist-speak for a snail, and one of the neatest groups that ever evolved.

General circulation model (GCM): A complex, three-dimensional climate simulation generally run on supercomputer.

Genus: A category in the taxonomic hierarchy between family and species.

Gondwana: The southern supercontinent including South America, Africa, India, and Antarctica.

Guild: An ecological association based on shared modes of life (e.g., sessile filter-feeders) rather than evolutionary descent.

Impact hypothesis: A theory suggesting that mass extinctions were caused by the impact of an extraterrestrial object and the subsequent global cooling.

Infaunal: Organisms that live within the sediment.

Isochronous: Two or more events happening at the same time (within the limits of resolution).

Isotope: One of several varieties of an element that differ in the number of neutrons.

Lazarus taxa: Groups that disappear from the fossil record for some period of time, generally during a mass extinction, only to reappear later in the fossil record.

Limestone: A carbonate rock composed of calcite or aragonite.

Magnetic reversal: A change in the polarity of the earth's magnetic field.

Mantle: The portion of the earth from 2,900 to 40 kilometers, largely composed of silicate minerals.

Mass extinction: An event in which a large number of organisms in many different clades are eliminated.

Mesozoic: The interval of geologic time from the Triassic through the Cretaceous Periods, roughly 250 to 65 million years ago.

Modern Evolutionary Fauna: As defined by Sepkoski (1984), the classes of marine taxa that reach the peak of their diversity during the post-Paleozoic, including gastropods, bivalves, malacostracan arthropods, bony fish, etc.

Molluscs: A phylum of invertebrates including snails, clams, and octopus.

Monophyletic: A group of organisms including the ancestral species and all descendent species.

Orogeny: A period of mountain building.

Paleozoic: The interval of geologic time from the Cambrian through the Permian Periods, roughly 540 to 245 million years ago.

Paleozoic Evolutionary Fauna: As defined by Sepkoski (1984), including the classes that reach peak diversity during the Ordovician-Permian, including articulate brachiopods, crinoids, etc.

Pangea: The late Paleozoic-Mesozoic supercontinent that included most present-day continental masses.

Paraphyletic: A clade of organisms in which either the ancestral species or some descendent species have been removed by systematists to another group, usually because of morphologic distinctiveness.

Period: One of the major temporal divisions of the geologic time scale corresponding to a system on the time-stratigraphic scale.

Periodicity hypothesis: The proposal that mass extinctions have occurred approximately every 26.4 million years over the past 250 million years.

Phylum: The broadest level of the taxonomic hierarchy; above the level of class.

Plate tectonics: A theory of global dynamics involving the movement of rigid surficial plates.

Pseudoextinction: The apparent extinction of a taxon due to taxonomic practice, i.e., renaming a species.

Province: A large biogeographic unit generally defined on the basis of common species that are distinct from adjacent regions. There are about 30 marine provinces today.

Pyroclastic eruption: An explosive volcanic eruption generally producing a tall column of ash.

Reefs: A solid limestone structure constructed by organisms and resistant to wave activity.

Regression: Migration of shoreline and associated environments toward the sea.

Series: A time-stratigraphic geologic unit below a system and above a stage.

Sessile: Organisms living fixed in one place and unable to move.

Species: A group of organisms capable of interbreeding and producing viable offspring and sharing a common evolutionary history.

Species-area effect: The claim that biologic diversity is controlled by a linear relationship between the amount of available area and diversity.

Stratigraphic range: The temporal duration of a species or other taxon. The recorded stratigraphic range is almost always less than the true stratigraphic range because of preservation problems.

Superchron: A lengthy interval of stable magnetic polarity, either normal or reversed.

Stable isotopes: Different isotopes of an element that are not subject to radioactive decay.

Stage: The time-stratigraphic unit shorter than a period and a series. Often the shortest interval that may be reliably correlated around the world.

Suspension feeder: A marine organism that feeds by straining organic material from water.

System: The rock deposited during a particular geologic period; the Permian System includes all sediment deposited during the Permian Period.

Taphonomy: The branch of paleontology dealing with the death, decay, and preservation of organisms.

Taxon: A named group of organisms within the Linnaen hierarchy of species, genus, family, etc.

Taxonomy: The science of the systematic relationships of plants and animals.

Transgression: A rise in relative sea level.

Trophic role: An organism's or species' mode of gathering food. For example, filter-feeding, predation, etc.

Type section: An internationally agreed-upon outcrop that serves as a reference section and defines the position of a geologic boundary.

Unconformity: A gap or break in the stratigraphic sequence due to either nondeposition or erosion.

Upwelling: Movement of dense, cold, nutrient-rich water up from ocean depths to the surface, enhancing primary productivity.

REFERENCES

Abich, H. W. 1878. *Geologische Forschungen in den Kaukasischen Ländern. Eine Bergkalkfauna aus der Araxes-Enge bei Djoulfa in Armenia.* Vienna.

Alekseev, A. S., L. D. Barsokova, G. M. Koesov, M. A. Nazarov, and A. G. Grigoryan. 1983. Permian-Triassic boundary event: geochemical investigations of the Transcaucasia section. *Abstracts of 14th Lunar and Planetary Science Conference, March 14–18,* pp. 3–4.

Allmon, W. D., D. H. Erwin, R. M. Linsley, and P. J. Morris. Forthcoming. Trophic level and evolutionary rate in Paleozoic gastropoda.

Alt, D., J. W. Sears, and D. W. Hyndman. 1988. Terrestrial maria; the origins of large basalt plateaus, hotspot tracks and spreading ridges. *Journal of Geology* 96:647–662.

Alvarez, L. W., W. Alvarez, F. Asaro, and H. V. Michel. 1980. Extra-terrestrial cause for the Cretaceous-Tertiary extinction. *Science* 208:1094–1108.

Anderson, D. L. 1982. Hotspots, polar wander, Mesozoic convection and the geoid. *Nature* 297:391–393.

Anderson, J. M. 1981. World Permo-Triassic correlations: their biostratigraphic basis. In M. M. Cresswell and P. Vella, eds., *Gondwana Five,* pp. 3–10. Rotterdam: A. A. Balkema.

REFERENCES

Anonymous. 1992. *Volcanism and Climate Change*. Washington, D.C.: American Geophysical Union.

Aplonov, S. 1988. An aborted Triassic ocean in west Siberia. *Tectonics* 7:1103–1122.

Asaro, F., L. W. Alvarez, W. Alvarez, and H. V. Michel. 1982. Geochemical anomalies near the Eocene/Oligocene and Permian/Triassic boundaries. In L. T. Silver and P. H. Schultz, eds., *Geological Implications of Impacts of Large Asteroids and Comets on the Earth*, Special Paper 190, pp. 517–528. Boulder, Colo.: Geological Society of America.

Assereto, R., A. Bosellini, N. Fantini Sestini, and W. C. Sweet. 1973. The Permian-Triassic boundary in the southern Alps (Italy). In A. Logan and L. V. Hills, eds. *The Permian and Triassic Systems and Their Mutual Boundary* Memoir 2, pp. 176–199. Calgary: Canadian Society of Petroleum Geologists.

Attrep, M., Jr., C. J. Orth, and L. R. Quintana. 1991. The Permian-Triassic of the Gartnerkofel-1 Core (Carnic Alps, Austria): geochemistry of common and trace elements II—INAA and RNAA. *Abhandlungen der Geologischen Bundesanstalt* 45:123–138.

Audley-Charles, M. G., and A. Hallam, eds. 1988. *Gondwana and Tethys*. Oxford: Geological Society of London.

Baksi, A. K. 1990. Search for periodicity in global events in the geologic record: Quo vadimus?. *Geology* 18:983–986.

Baksi, A. K., and E. Farrar. 1991. $^{40}Ar/^{39}Ar$ dating of the Siberian traps, USSR: evaluation of the ages of the two major extinction events relative to episodes of flood-basalt volcanism in the USSR and the Deccan Traps, India. *Geology* 19:461–464.

Balme, B. E., and R. J. Helby. 1973. Floral modifications at the Permian-Triassic boundary in Australia. In A. Logan and L. V. Hills, eds. *The Permian and Triassic Systems and Their Mutual Boundary* Memoir 2, pp. 433–444. Calgary: Canadian Society of Petroleum Geologists.

Bambach, R. K. 1977. Species richness in marine benthic habitats through the Phanerozoic. *Paleobiology* 3:152–167.

Bambach, R. K. 1983. Ecospace utilization and guilds in marine communities through the Phanerozoic. In M. J. S. Tevesz and P. L. McCall, eds., *Biotic Interactions in Recent and Fossil Benthic Communities*, pp. 719–746. New York: Plenum Press.

Bambach, R. K. 1985. Classes and adaptive variety: the ecology of diversification in marine faunas through the Phanerozoic. In J. W. Valentine, ed., *Phanerozoic Diversity Patterns*, pp. 191–253. Princeton: Princeton University Press.

Bambach, R. K. 1989. Similarities and differences in diversity patterns at different taxonomic levels using traditional (non-cladistic) groupings. *Geological Society of America Abstracts with Program* 21(6):206–207.

Bambach, R. K. 1990. Late Paleozoic provinciality in the marine realm. In W.

S. McKerrow and C. R. Scotese, eds., *Palaeozoic Palaeogeography and Biogeography*, pp. 307–323. London: Geological Society Memoir.

Bambach, R. K. 1992. Personal communication.

Bando, Y. 1979. Upper Permian and Lower Triassic ammonoids from Abadeh, Central Iran. *Memoirs of the Faculty of Education, Kagawa University, Part II* 29:103–138.

Bando, Y. 1980. On the Otoceratacean ammonoids in the Central Tethys with a note on their evolution and migration. *Memoirs of the Faculty of Education, Kagawa University, Part II* 30:23–49.

Barley, M. E., and D. I. Groves. 1992. Supercontinent cycles and the distribution of metal deposits through time. *Geology* 20:291–294.

Banks, P. O. 1973. Permian-Triassic radiometric time scale. In A. Logan and L. V. Hills, eds., *The Permian and Triassic Systems and Their Mutual Boundary* Memoir 2, pp. 669–677. Calgary: Canadian Society of Petroleum Geologists.

Bassett, D. A. 1991. Roderick Murchison's *The Silurian System*: a sesquicentennial tribute. *Special Papers in Palaeontology* 44:7–90.

Batten, R. L. 1973. The vicissitudes of the gastropods during the interval of Guadalupian-Ladinian time. In A. Logan, and L. V. Hills, eds., *The Permian and Triassic Systems and Their Mutual Boundary*, Memoir 2, pp. 596–607. Calgary: Canadian Society of Petroleum Geologists.

Baud, A., M. Magaritz, and W. T. Holser. 1989. Permian-Triassic of the Tethys: carbon isotope studies. *Geologische Rundschau* 78:649–677.

Benson, R. H. 1984. The Phanerozoic "crisis" as viewed from the Miocene. In W. A. Berggren and J. A. van Couvering, eds., *Catastrophes and Earth History*, pp. 437–446. Princeton: Princeton University Press.

Benton, M. J. 1985. Mass extinction among non-marine tetrapods. *Nature* 316:811–814.

Benton, M. J. 1987a. The history of the biosphere: equilibrium and non-equilibrium models of global diversity. *Trends in Ecology and Evolution* 2:153–156.

Benton, M. 1987b. Mass extinctions among families of non-marine tetrapods: the data. *Mémoir. Société Géologique France* 150:21–32.

Benton, M. J. 1987c. Progress and competition in macroevolution. *Biological Reviews* 62:305–338.

Benton, M. J. 1988. Mass extinctions in the fossil record of reptiles: paraphyly, patchiness and periodicity (?). In G. P. Larwood, ed., *Extinction and Survival in the Fossil Record*, pp. 269–294. Oxford: Clarendon Press.

Benton, M. J. 1990a. The causes of the diversification of life. In P.D. Taylor and G. P. Larwood, eds., *Major Evolutionary Radiations*, pp. 409–430. Oxford: Clarendon Press.

Benton, M. J. 1990b. *Vertebrate Palaeontology*. London: Unwin Hyman.

Berner, R. A. 1989a. Biogeochemical cycles of carbon and sulfur and their effect on atmospheric oxygen over Phanerozoic time. *Palaeogeography,*

Palaeoclimatology, Palaeoecology (Global and Planetary Change Section) 75:97–122.

Berner, R. A. 1989b. Dying, O_2 and mass extinction. *Nature* 340:603–604.

Berner, R. A. 1990. Atmospheric carbon dioxide levels over Phanerozoic time. *Science* 249:1382–1386.

Berner, R. A. 1991. A model for atmospheric CO_2 over Phanerozoic time. *American Journal of Science* 291:339–376.

Berner, R. A., and A. C. Lasaga. 1989. Modeling the geochemical carbon cycle. *Scientific American* March:73–81.

Berner, R. A., and R. Raiswell. 1983. Burial of organic carbon and pyrite sulfur in sediments over Phanerozoic time. *Geochimica et Cosmochimica Acta* 47:855–862.

Berry, W. B. N., and P. Wilde. 1978. Progressive ventilation of the oceans—an explanation for the distribution of the Lower Paleozoic black shales. *American Journal of Science* 278:257–275.

Beurlen, K. 1956. Der Faunenschnitt an der Perm-Trias Grenze. *Zeitschrift der Deutschen Geologischen Gesellschaft* 108:88–99.

Bice, D. M., C. R. Newton, S. McCauley, P. W. Reiners, and C. A. McRoberts. 1992. Shocked quartz at the Triassic-Jurassic boundary in Italy. *Science* 255:443–446.

Bonan, G. B., D. Pollard, and S. L. Thompson. 1992. Effects of boreal forest vegetation on global climate. *Nature* 359:716–718.

Bottjer, D. J., and W. I. Ausich. 1986. Phanerozoic development of tiering in soft substrata suspension-feeding communities. *Paleobiology* 12:400–420.

Bottjer, D. J., M. L. Droser, and C. Y. Wang. 1988. Fine-scale resolution of mass extinction events: trace fossil evidence from the Permian-Triassic boundary in South China. *Geological Society of America Abstracts with Program* 20:A106

Bowen, R. L. 1968. Paleoclimatic and paleobiologic implications of Louann slat deposition. *American Association of Petroleum Geologists Bulletin* 52:1833.

Bramlette, M. N. 1965. Massive extinctions in biota at the end of Mesozoic time. *Science* 148:1696–1699.

Brasier, M. D. 1988. Foraminiferid extinction and ecological collapse during global biological events. In G. P. Larwood, ed., *Extinction and Survival in the Fossil Record*, pp. 37–64. Oxford: Oxford University Press.

Briggs, D. E. G., R. A. Fortey, and E. N. K. Clarkson. 1988. Extinction and the fossil record of the arthropods. In G. P. Larwood, ed. *Extinction and Survival in the Fossil Record*. pp.171–209. Oxford: Oxford University Press.

Broglio Loriga, C., C. Neri, M. Pasini, and R. Posenato. 1988. Marine fossil assemblages from Upper Permian to lowermost Triassic in the western Dolomites (Italy). *Memorie della Società Geologica Italiana* 34:5–44.

Broglio-Loriga, C., and G. Cassinis. 1992. The Permo-Triassic boundary in the Southern Alps (Italy) and in adjacent Periadratic regions. In W. C. Sweet,

Z. Y. Yang, J. M. Dickins, and H. F. Yin, eds., *Permo-Triassic Events in the Eastern Tethys*, pp. 78–97. Cambridge: Cambridge University Press.

Buggisch, W., and S. Noè. 1988. Upper Permian and Permian-Triassic boundary of the Carnia (Bellerophon Formation, Tesero Horizon, Northern Italy). *Memorie della Società Geologica Italiana* 34:91–106.

Buser, S., K. Grad, B. Ogorelec, A. Ramovs, and L. Sribar. 1988. Stratigraphic, paleontological and sedimentologic characteristics of Upper Permian beds in Slovenia, NW Yugoslavia. *Memorie della Società Geologica Italiana* 34:195–210.

Caldiera, K., and M. R. Rampino. 1990a. Carbon dioxide emissions from Deccan volcanism and a K/T boundary greenhouse effect. *Geophysical Research Letters* 17:1299–1302.

Caldeira, K. G., and M. R. Rampino. 1990b. Deccan volcanism, greenhouse warming, and the Cretaceous/Tertiary boundary. In V. L. Sharpton and P. D. Ward, eds., *Global Catastrophes in Earth History*. Geological Society of America Special Paper 247, pp. 117–123. Boulder, Colo.: Geological Society of America.

Caldeira, K., and M. R. Rampino. 1991. The mid-Cretaceous superplume, carbon dioxide, and global warming. *Geophysical Research Letters* 18:987–990.

Campbell, I. H., G. K. Czamanski, V. A. Fedorenko, R. I. Hill, and V. Stepanov. 1992. Synchronism of the Siberian Traps and the Permian-Triassic boundary. *Science* 258:1760–1763.

Carey, S. W. 1983. Tethys, and her forebears. In S. W. Carey, ed., *The Expanding Earth Symposium*, pp. 169–187. Sydney, Australia: University of Tasmania.

Carlson, S. J. 1991. A phylogenetic perspective on articulate brachiopod diversity and the Permo-Triassic extinction. In E. Dudley, ed., *The Unity of Evolutionary Biology*. Proceedings of the 4th International Congress of Systematic and Evolutionary Biology, pp. 119–142. Portland, Ore.: Dioscorides Press.

Carpenter, F. M., and L. Burnham. 1985. The geological record of insects. *Annual Review of Earth and Planetary Science* 13:297–314.

Carr, T. R., and J. A. Kitchell. 1980. Dynamics of taxonomic diversity. *Paleobiology* 6:427–443.

Cassinis, G., ed. 1988. Proceedings of the Field Conference on: Permian and Permian-Triassic boundary in the south-Alpine segment of the Western Tethys, and additional regional reports. *Memorie della Società Geologica Italiana* 34:1–366.

Cathles, L. M., and A. Hallam. 1991. Stress-induced changes in plate density, Vail sequences, epeirogeny and short-lived global sea level fluctuations. *Tectonics* 10:659–671.

Chaloner, W. G., and S. V. Meyen. 1973. Carboniferous and Permian floras of the northern continents. In A. Hallam, ed., *Atlas of Palaeobiogeography*, pp. 69–186. Amsterdam: Elsevier.

REFERENCES

Chao, K. K. 1965. The Permian ammonoid-bearing faunas of South China. *Scientia Sinica* 14:1813–1825.

Chase, C. G. 1979. Subduction, the geoid, and lower mantle convection. *Nature* 282:464–468.

Chase, C. G. 1985. The geological significance of the geoid. *Annual Review of Earth and Planetary Sciences* 13:97–117.

Chen, J., M. Shao, W. Huo, and Y. Yao. 1984. Carbon isotope of carbonate strata at Permian-Triassic boundary in Changxing, Zhejiang. *Acta Geologica Sinica* 1984:88–93.

Chengfa, C. F., N. S. Chen, D. Wanming, J. F. Dewey, A. Gansser, N. B. W. Harris, C. W. Jin, W. S. F. Kidd, M. R. Leeder, H. Li, C. J. Liu, H. J. Mei, P. Molnar, Y. Pan, J. A. Pearce, R. M. Shackleton, A. B. Smith, Y. Y. Sun, M. Ward, D. R. Watts, J. T. Xu, R. H. Xu, J. X. Yin, and Y. Q. Zhang. 1986. Preliminary conclusions of the Royal Society and Academia Sinica 1985 geotraverse of Tibet. *Nature* 323:501–507.

Chesser, C. A., W. I. Rose, A. Deino, R. Drake, and J. A. Westgate. 1991. Eruptive history of earth's largest quaternary caldera (Toba, Indonesia) clarified. *Geology* 19:200–203.

Christiansen, R. L. 1979. Cooling units and composite sheets in relation to caldera structure. In C. E. Chapin and W. E. Easton, eds. *Ash Flow Magmatism*, Special Paper 180, pp. 5–27. Boulder, Colo.: Geological Society of America.

Christiansen, R. L. 1984. Yellowstone magmatic evolution: its bearing on understanding large-volume explosive volcanism. In *Explosive Volcanism: Inception, Evolution, and Hazards*, pp. 85–95. Washington, D.C.: National Academy of Sciences.

Claoue-Long, J. C., Z. C. Zhang, G. G. Ma, and S. H. Du. 1991. The age of the Permian-Triassic boundary. *Earth and Planetary Science Letters* 105:182–190.

Clark, D. J., C.-Y. Wang, C. J. Orth, and J. S. Gilmore. 1986. Conodont survival and low iridium abundances across the Permian-Triassic boundary in South China. *Science* 233:984–986.

Clark, D. L. 1987. Conodonts: the final fifty million years. In R. J. Aldridge, ed., *Palaeobiology of Conodonts*, pp. 165–174. Chichester: Ellis Horwood Ltd.

Claypool, G. E., W. T. Holser, I. R. Kaplan, H. Sakai, and I. Zak. 1980. The age curves of sulfur and oxygen isotopes in marine sulfate and their mutual interpretation. *Chemical Geology* 28:199–260.

Colbert, E. N. 1973. Tetrapods and the Permian-Triassic transition. In A. Logan and L. V. Hills, eds. *The Permian and Triassic Systems and Their Mutual Boundary* Memoir 2, pp.481–492. Calgary: Canadian Society of Petroleum Geologists.

Colbert, E. N. 1986. Therapsids in Pangea and their contemporaries and competitors. In N. Hotton , P. D. MacLean , J. J. Roth, and E. C. Roth, eds.,

Ecology and Biology of Mammal-Like Reptiles, pp. 133–145. Washington,D.C.: Smithsonian Institution Press.

Conner, E. F., and E. D. McCoy. 1979. The statistics and biology of the species-area relationship. *American Naturalist* 113:791–833.

Cooper, G. A., and R. E. Grant. 1972. Permian brachiopods of West Texas, I. *Smithsonian Contributions to Paleobiology*, No. 14.

Courtillot, V. 1990. Deccan volcanism at the Cretaceous-Tertiary boundary: past climatic crises as a key to the future? *Palaeogeography, Palaeoclimatology, Palaeoecology (Global and Planetary Change Section)* 189:291–299.

Courtillot, V., and J. Besse. 1987. Magnetic field reversals, polar wander and core-mantle coupling. *Science* 237:1140–1147.

Cox, C. B. 1967. Changes in terrestrial vertebrate faunas during the Mesozoic. In W. B. Harland, C. H. Holland, M. R. House, N. F. Hughes, A. B. Reynolds, M. J. S. Rudwick, G. E. Stratterwaite, L. B. H. Tarlo, and E. C. Willey, eds. *The Fossil Record*, pp. 77–89. London: Geological Society.

Cox, K. G. 1991. A superplume in the mantle. *Nature* 352:564–565.

Cracraft, J. 1985. Biological diversification and its causes. *Annals of the Missouri Botanical Garden* 72:794–822.

Crowell, J. C. 1978. Gondwanan glaciation, cyclothems, continental positioning and climate change. *American Journal of Science* 278:1345–1372.

Crowell, J. C. 1983. Ice ages recorded on Gondwanan continents. *Geological Society of South Africa Transactions* 86:237–262.

Crowell, J. C. Forthcoming. The ending of the Late Paleozoic Ice Age during the Permian Period. In P. Scholle, ed. *The Permian of the Northern Continents*. Berlin: Springer-Verlag.

Crowley, T. J., and S. K. Baum. 1991. Estimating Carboniferous sea-level fluctuations from Gondwanan ice extent. *Geology* 19:975–977.

Crowley, T. J., and S. K. Baum. 1992. Modeling late Paleozoic glaciation. *Geology* 20:507–510.

Crowley, T. J., W. T. Hyde, and D. A. Short. 1989. Seasonal cycle variations on the supercontinent of Pangea. *Geology* 17:457–460.

Crowley, T. J., J. G. Mengel, and D. A. Short. 1987. Gondwanaland's seasonal cycle. *Nature* 329:803–807.

Crowley, T. J., and G. R. North. 1991. *Paleoclimatology*. New York: Oxford University Press.

Dalziel, I. W. D. 1992. On the organization of American plates in the Neoproterozoic and the breakout of Laurentia. *GSA Today* 2:237, 240–241.

Demenitskaya, R. M., and S. V. Aplonov. 1988. The demise of small oceans in the geologic past. *International Geological Review* 30:1043–1051.

Desmond, A. 1982. *Archetypes and Ancestors: Paleontology in Victorian London 1850–1875*. Chicago: University of Chicago Press.

Dickins, J. M. 1983. Permian to Triassic changes in life. *Memoir of the Australasian Paleontologists* 1:297–303.

REFERENCES

Dickins, J. M. 1985. Late Paleozoic glaciation. *Journal of Australian Geology and Geophysics* 9:163–169.

Diener, K. 1912. The Trias of the Himalayas. *Geological Survey of India, Memoir* 36:202–360.

DiMichele, W. A. 1992. Personal communication.

DiMichele, W. A., and R. B. Aronson. 1992. The Pennsylvanian-Permian vegetational transition: a terrestrial analogue to the onshore-offshore hypothesis. *Evolution* 46:807–824.

DiMichele, W. A., and R. W. Hook. 1992. Paleozoic terrestrial ecosystems. In A. K. Behrensmeyer, J. D. Danuth, W. A. DiMichele, R. Potts, H.-D. Sues, and S. Wing, eds., *Evolutionary Paleoecology of Terrestrial Plants and Animals*, pp. 205–325. Chicago: University of Chicago Press.

Donovan, S. T., and E. J. W. Jones. 1979. Causes of world-wide changes in sea-level. *Journal of the Geological Society of London* 136:187–194.

Dunbar, C. O. 1955. Permian brachiopod faunas of central East Greenland. *Meddelesler om Grønland* 110:1–169.

Enegebretson, D. C., K. P. Kelley, H. J. Cashman, and M. A. Richards. 1992. 180 million years of subduction. *GSA Today* 2:93–95, 100.

Engel, A. E. J., and C. G. Engel. 1964. Continental accretion and the evolution of North America. In A. P. Subramaniam and S. Balakrishna, eds. *Advancing Frountiers in Geology and Geophysics*, pp. 17–37. Hyderabad: Indian Geophysical Union.

Erwin, D. H. 1985. The Cerithiacea, Subulitacea, Pyramidellacea and Acteonacea of the Permian Basin, West Texas and New Mexico with a consideration of Permo-Triassic Gastropod Dynamics. Doctoral dissertation, University of California, Santa Barbara.

Erwin, D. H. 1989a. The end-Permian mass extinction: what really happened and did it matter? *Trends in Ecology and Evolution* 4:225–229.

Erwin, D. H. 1989b. Regional paleoecology of Permian gastropod genera, southwestern United States and the end-Permian mass extinction. *Palaios* 4:424–438.

Erwin, D. H. 1990a. Carboniferous-Triassic gastropod diversity patterns and the Permo-Triassic mass extinction. *Paleobiology* 16:187–203.

Erwin, D.H. 1990b. The end-Permian mass extinction. *Annual Review of Ecology and Systematics* 21:69–91.

Erwin, D. H. 1991. A preliminary classification of evolutionary radiations. *Historical Biology* 6:133–147.

Erwin, D. H. Forthcoming. The origin of metazoan development: a paleobiological perspective. *Biological Journal of the Linnean Society*.

Erwin, D. H., and J. W. Valentine, J. W. 1984. Differential gastropod extinctions: possible role of larval strategies. *Geological Society of America, Abstracts with Program* 16(6):503.

Erwin, D. H., J. W. Valentine, and J. J. Sepkoski, Jr. 1987. A comparative study of diversification events: the early Paleozoic versus the Mesozoic. *Evolution* 41:1177–1186.

Erwin, D. H., and T. A. Vogel. 1992. Testing for causal relationships between large pyroclastic volcanic eruptions and mass extinctions. *Geophysical Research Letters* 19:893–896.

Eshet, Y. 1990. Paleozoic-Mesozoic palynology of Israel. I. Palynological aspects of the Permian-Triassic succession in the subsurface of Israel. *Geological Survey of Israel, Bulletin* 81:1–57.

Eshet, Y. 1992. The palynofloral succession and palynological events in the Permo-Triassic boundary interval in Israel. In W. C. Sweet, Z. Y. Yang, J. M. Dickins, and H. F. Yin, eds., *Permo-Triassic Boundary Events in the Eastern Tethys*, pp. 134–145. Cambridge: Cambridge University Press.

Evans, J. 1874. The anniversary address of the president. *Quarterly Journal of the Geological Society* 31:xxxvii-xliii.

Fedorowski, J. 1989. Extinction of Rugosa and Tabulata near the Permian/Triassic boundary. *Acta Palaeontologica Polonica* 34:47–70.

Fischer, A. G. 1965. Brackish oceans as the cause of the Permo-Triassic marine faunal crisis. In A. E. M. Nairn, ed., *Problems in Palaeoclimatology*, pp. 566–574. London: Interscience.

Fischer, A. G. 1981. Climatic oscillations in the biosphere. In M. H. Nitecki, ed., *Biotic Crises in Ecological and Evolutionary Time*, pp. 103–131. New York: Academic Press.

Fischer, A. G. 1984. The two Phanerozoic supercycles. In W. A. Berggren, and J. A. Van Couvering, eds., *Catastrophes and Earth History*, pp. 129–150. Princeton: Princeton University Press.

Fischer, A. G., and M. A. Arthur. 1977. Secular variations in the pelagic realm. In H. E. Cook and P. Enos, eds., *Deep Water Carbonate Environments*, pp. 18–50. Tulsa: SEPM Special Publication No. 25.

Flessa, K. W., and D. Jablonski. 1985. Declining Phanerozoic background extinction rates: effect of taxonomic structure? *Nature* 313:216–218.

Flessa, K. W., and J. J. Sepkoski. 1978. On the relationship between Phanerozic diversity and changes in habitable area. *Paleobiology* 4:359–366.

Flügel, E., and J. Reinhardt. 1989. Uppermost Permian reefs in Skyros (Greece) and Sichuan (China): implications for the Late Permian Extinction Event. *Palaios* 4:502–518.

Flügel, E., and G. D. Stanley. 1984. Reorganization, development and evolution of post-Permian reefs and reef organisms. *Palaeontographica Americana* 54:177–186.

Forney, G. G. 1975. Permo-Triassic sea level change. *Journal of Geology* 83:773–779.

Fortey, R. A., and R. M. Owens. 1990. Trilobites. In K. J. McNamara, ed., *Evolutionary Trends*, pp. 121–142. Tuscon: University of Arizona Press.

Frakes, L. A., and J. E. Francis. 1988. A guide to Phanerozoic cold polar climates from high-latitude ice-rafting in the Cretaceous. *Nature* 333:547–549.

Frederiksen, N. O. 1972. The rise of the Mesophytic flora. *Geoscience and Man* 4:17–28.

REFERENCES

Fu, S. L. 1988. Palaeontologic characters of the Dalong Formation of Jiang-shan, Zhejiang and the boundary between the Permian and Triassic. *Journal of Stratigraphy* 12(2):136–142.

Furnish, W. M., and B. F. Glenister. 1970. Permian ammonoid *Cyclolobus* from the Salt Range, West Pakistan. In B. Kummel and C. Teichert, eds., *Stratigraphic Boundary Problems: Permian and Triassic of West Pakistan*, pp. 153–175. University of Kansas Department of Geology Special Publication No. 4. Lawrence: University Press of Kansas.

Gale, A. S. 1987. Phylogeny and classification of the Asteroidea (Echinodermata). *Zoological Journal of the Linnean Society* 89:107–132.

Golshani, F., H. Partozar, and K. Seyed-Emami. 1988. Permian-Triassic boundary in Iran. *Memorie della Società Geologica Italiana* 34:257–262.

Gordon, W. A. 1975. Distribution by latitude of Phanerozoic evaporite deposits. *Journal of Geology* 83:671–684.

Gould, S. J. 1985. The paradox of the first tier: an agenda for paleobiology. *Paleobiology* 11:2–12.

Gould, S. J., and C. B. Calloway. 1980. Clams and brachiopods—ships that pass in the night. *Paleobiology* 6:383–396.

Grant, R. E. 1970. Brachiopods from Permian-Triassic boundary beds and age of Chhidru Formation, West Pakistan. *University of Kansas, Department of Geology Special Publication* 4:117–153.

Grant, R. E., M. K. Nestell, A. Baud, and C. Jenny. 1991. Permian stratigraphy of Hydra Island, Greece. *Palios* 6:479–497.

Gregory, W. K. 1952. *Evolution Emerging*, vol. II. New York: Macmillan.

Griffin, S. R. 1992. Comment and Reply on "Latest pulse of Earth: Evidence for a mid-Cretaceous superplume" and "Geological consequences of superplumes." *Geology* 20:475–477.

Gruszczyński, M., S. Halas, A. Hoffman, and K. Małkowski. 1989. A brachiopod calcite record of the oceanic carbon and oxygen isotope shifts at the Permian/Triassic transition. *Nature* 337:64–68.

Gruszczyński, M., S. Hoffman, K. Małkowski, and J. Veizer. 1992. Seawater strontium isotopic perturbation at the Permian-Triassic boundary, west Spitsbergen, and its implications for the interpretation of strontium isotopic data. *Geology* 20:779–782.

Gruszczyński, M., A. Hoffman, K. Małkowski, K. Zawidzka, S. Halas, and Y. Zeng. 1990. Carbon isotope drop across the Permian-Triassic boundary in SE Sichuan, China. *Neues Jahrbuch fur Geologische und Palaontologische Monatshaft* 10:600–606.

Gurnis, M. 1988. Large-scale mantle convection and the aggregation or dispersal of supercontinents. *Nature* 332:695–699.

Gurnis, M. 1990. Ridge spreading, subduction, and sea level. *Science* 250:970–972.

Gurnis, M. 1992a. Long-term controls on eustatic and epirogenic motions by mantle convection. *GSA Today* 2:142, 144–145, 157.

Gurnis. M. 1992b. Rapid continental subsidence following the initiation and evolution of subduction. *Science* 255:1556–1558.

Haag, M., and F. Heller. 1991. Late Permian to Early Triassic magnetostratigraphy. *Earth and Planetary Science Letters* 107:42–54.

Haas, J., F. Góczán, A. Oravecz-Scheffer, A. Barabas-Sthul, G. Majoros, and A. Berczi-Makk. 1988. Permian-Triassic boundary in Hungary. *Memorie della Società Geologica Italiana* 34:221–241.

Hallam, A. 1971. Re-evaluation of the palaeogeographic argument for an expanding earth. *Nature* 232:180–182.

Hallam, A. 1977. Secular changes in marine inundation of USSR and North America through the Phanerozoic. *Nature* 269:769–772.

Hallam, A. 1983. Supposed Permo-Triassic megashear between Laurasia and Gondwana. *Nature* 301:499–502.

Hallam, A. 1984a. Pre-Quaternary sea-level changes. *Annual Review of Earth and Planetary Science* 12:205–243.

Hallam, A. 1984b. The unlikelihood of an expanding Earth (book review). *Geological Magazine* 121:653–655.

Hallam, A. 1989a. The case for sea-level change as a dominant causal factor in mass extinctions of marine invertebrates. *Philosophical Transactions of the Royal Society, London, Ser. B* 325:437–455.

Hallam, A. 1989b. *Great Geological Controversies.* Oxford: Oxford University Press.

Hallam, A. 1991. Why was there a delayed radiation after the end-Paleozoic extinctions? *Historical Biology* 5:257–262.

Hallam, A. 1992. *Phanerozoic Sea-Level Changes.* New York: Columbia University Press.

Hallam, A., and A. I. Miller, 1988. Extinction and survival in the Bivalvia. In G. P. Larwood, ed., *Extinction and Survival in the Fossil Record,* pp. 121–138. Oxford: Clarendon Press.

Hansen, T. A. 1987. Extinction of late Eocene to Oligocene molluscs: relationship to shelf area, temperature changes and impact events. *Palaios* 2:69–75.

Harland, W. B., R. L. Armstrong, A. V. Cox, L. E. Craig, A. G. Smith, and D. G. Smith. 1989. *A Geologic Time Scale 1989.* Cambridge: Cambridge University Press.

Hatfield, C. B., and M. J. Camp. 1970. Mass extinctions correlated with periodic galactic events. *Geological Society of America Bulletin* 81:911–914.

Hay, W. W., and J. R. Southam. 1977. Modulation of marine sedimentation by the continental shelves. In N. R. Anderson and A. Malahoff, eds., *The Role of Fossil Fuel CO_2 in the Oceans,* pp. 569–605. New York: Plenum.

Hays, J. D., and W. C. Pitman. 1973. Lithospheric plate motion, sea level changes and climatic and ecological consequences. *Nature* 246:18–22.

He, J. W., L. Rui, C. F. Chai, and S. I. Ma. 1987. The latest Permian and earliest Triassic volcanic activities in the Meishan area of Chanxing, Zhejiang. *Journal of Stratigraphy* 11(3):194–199.

REFERENCES

Heller, F., W. Lowrie, H. Li, and J. Wang. 1988. Magnetostratigraphy of the Permo-Triassic boundary section and Shangsi (Guangyan, Sichuan Province, China). *Earth and Planetary Science Letters* 88:348–356.

Hiller, N., and N. Stavrakis. 1984. Permo-Triassic fluvial systems in the southeastern Karoo Basin, South Africa. *Palaeogeography, Palaeoclimatology, Palaeoecology* 45:1–21.

Hirsch, F., and T. Weissbrod. 1988. The Permian-Triassic boundary in Israel. *Memorie della Società Geologica Italiana* 34:253–256.

Hoffman, A. 1985a. Island biogeography and paleobiology: in search of evolutionary equilibria. *Biological Review* 60:455–471.

Hoffman, A. 1985b. Patterns of family extinction depend on definition and geological timescale. *Nature* 315:659–662.

Hoffman, A. 1986. Neutral model of Phanerozoic diversification: Implications for macroevolution. *Neues Jahrbuch fur Geologische und Palaontologische Abhandleungen* 172:219–244.

Hoffman, A. 1989a. *Arguments on Evolution*. New York: Oxfrod University Press.

Hoffman, A. 1989b. Changing paleontological views on mass extinction phenomena. In S. Donovan, ed., *Mass Extinctions: Processes and Evidence*, pp. 1–18. London: Belhaven Press.

Hoffman, A. 1989c. Mass extinctions: the view of a sceptic. *Journal of the Geological Society* 146:21–35.

Hoffman, A. 1989d. What, if anything, are mass extinctions? *Philosophical Transactions of the Royal Society, London. Ser. B* 325:253–262.

Hoffman, A., and J. Ghiold. 1985. Randomness in the pattern of "mass extinctions," and "waves of origination." *Geological Magazine* 122:1–4.

Hoffman, A., M. Gruszczyński, and K. Małkowski. 1990. Oceanic $\delta^{13}C$ values as indicators of atmospheric oxygen depletion. *Modern Geology* 14:211–221.

Hoffman, A., M. Gruszczyński, and K. Małkowski. 1991. On the interrelationship between temporal trends in $\delta^{13}C$, $\delta^{18}O$ and $\delta^{34}S$ in the world ocean. *Journal of Geology* 99:355–370.

Holland, H. D. 1984. *The Chemical Evolution of the Atmosphere and Oceans*. New York: Wiley.

Holmes, A. 1951. The sequence of Precambrian orogenic belts in south and central Africa. *18th International Geological Congress, London (1948)* 14:254–269.

Holser, W. T. 1984. Gradual and abrupt shifts in ocean chemistry during Phanerozoic time. In H. D. Holland, and A. F. Trendall, eds., *Patterns of change in Earth Evolution*, pp. 123–143. Berlin: Springer-Verlag.

Holser, W. T. 1992. Personal communication.

Holser, W. T., and M. Magaritz. 1987. Events near the Permian-Triassic boundary. *Modern Geology* 11:155–180.

Holser, W. T., and M. Magaritz. Forthcoming. Cretaceous/Tertiary and Permian/Triassic boundary events compared. *Geochimica et Cosmochimica Acta.*

Holser, W. T., M. Magaritz, and D. L. Clark. 1986. Carbon-isotope stratigraphic correlations in the Late Permian. *American Journal of Science* 286:390–402.

Holser, W. T., M. Magaritz, and J. Wright. 1986. Chemical and isotopic variations in the world ocean during Phanerozoic time. In O. Walliser, ed., *Global Bio-events*, pp. 63–74. Berlin: Springer-Verlag.

Holser, W. T., M. Schidlowski, F. T. MacKenzie, and J. B. Maynard. 1988. Geochemical cycles of carbon and sulfur. In C. B. Gregor, R. M. Garrels, F. T. MacKenzie, and J. B. Maynard, eds., *Chemical Cycles in the Evolution of the Earth*, pp. 105–173. New York: Wiley.

Holser, W. T., and H. P. Schönlaub. 1991. The Permian-Triassic boundary in the Carnic Alps of Austria (Gartnerkofel region). *Abhandlungen der Geologischen Bundesanstalt* 45:1–232.

Holser, W. T., H. P. Schönlaub, M. Attrep, Jr., K. Boeckelmann, P. Klein, M. Magaritz, C. J. Orth, A. Fenninger, C. Jenny, M. Kralik, H. Mauritsch, E. Pak, J. M. Schramm, K. Stattegger, and R. Schmoller. 1989. A unique geochemical record at the Permian/Triassic boundary. *Nature* 337:39–44.

Holser, W. T., H. P. Schönlaub, K. Boeckelmann, and M. Magaritz. 1991. The Permian-Triassic of the Gartnerkofel-1 Core (Carnic Alps, Austria): synthesis and conclusions. *Abhandlungen der Geologischen Bundesanstalt* 45:213–232.

Horita, J., T. J. Friedman, B. Lazar, and H. D. Holland. 1991. The composition of Permian seawater. *Geochimica et Cosmochimica Acta* 55:417–432.

Hotton, N., III. 1967. Stratigraphy and sedimentation in the Beaufort Series (Permian-Triassic), South Africa. In C. Teichert and E. L. Yochelson, eds., *Essays in Paleontology and Stratigraphy*, pp. 390–428. Lawrence: University of Kansas, Department of Geology Special Publication.

Hsu, K. J., H. Oberhansli, J. Y. Gao, S. Sun, H. Chen, and U. Krahenbuhl. 1985. "Strangelove ocean" before the Cambrian explosion. *Nature* 316:809–811.

Huang, K., N. D. Opdyke, X. Peng, and J. Li. 1992. Paleomagnetic results from the Upper Permian of the eastern Qiangtang Terrane of Tibet and their tectonic implications. *Earth and Planetary Science Letters* 111:1–10.

Huff, W. D., S. M. Bergstrom, and D. R. Kolata. 1992. Gigantic Ordovician volcanic ash fall in North America and Europe: Biological, tectonomagmatic, and event-stratigraphic significance. *Geology* 20:875–878.

Hüssner, M. 1983. Die Faunenwende Perm/Trias [The fauna of the Perm/Trias.] *Geologische Rundschau* 72:1–22.

Iranian-Japanese Research Group. 1981. The Permian and Lower Triassic systems in Abadeh Region, Central Iran. *Memoirs of the Faculty of Science, Kyoto University, Series in Geology and Mineralogy* 47:61–133.

Irving, E. 1977. Drift of major continental blocks since the Devonian. *Nature* 270:304–309.

REFERENCES

Irving, E., and G. Pullaiah. 1976. Reversals of the geomagnetic field, magnetostratigraphy, and relative magnitude of paleosecular variation in the Phanerozoic. *Earth Science Reviews* 12:35–64.

Jablonski, D. 1980. Apparent versus real effects of transgressions and regressions. *Paleobiology* 6:397–407.

Jablonski, D. 1985. Marine regressions and mass extinctions: a test using the modern biota. In J. W. Valentine, ed., *Phanerozoic Diversity Patterns*, pp. 335–354. Prineton: Princeton University Press.

Jablonski, D. 1986a. Background and mass extinctions: alternation of macroevolutionary regimes. *Science* 231:129–133.

Jablonski, D. 1986b. Causes and consequences of mass extinctions: a comparative approach. In D. K. Elliot, ed., *Dynamics of Extinction*, pp. 183–229. New York: John Wiley and Sons.

Jablonski, D. 1986c. Evolutionary consequences of mass extinction. In D. M. Raup, and D. Jablonski, eds., *Patterns and Processes in the History of Life*, pp. 313–329. Berlin: Springer-Verlag.

Jablonski, D. 1989. The biology of mass extinction: a palaeontological view. *Philosophical Transactions of the Royal Society, London Ser. B.* 325:357–368.

Jablonski, D., and D. J. Bottjer. 1990. The origin and diversification of major groups: environmental patterns and macroevolutionary lags. In P. D. Taylor and G. P. Larwood, eds., *Major Evolutionary Radiations*, pp. 17–57. Oxford: Clarendon Press.

Jablonski, D., and K. W. Flessa. 1986. The taxonomic structure of shallow-water marine faunas: implications for Phanerozoic extinctions. *Malacologia* 27:43–66.

James, N. P. 1983. Reefs. In P. A. Scholle, D. G. Begout, and C. H. Moore, eds., *Carbonate Depositional Environments*. American Association of Petroleum Geologists Memoir 33:345–462.

Johnson, J. G. 1971. Timing and coordination of orogenic, epeirogenic and eustatic events. *Geological Society of America Bulletin* 82:3263–3298.

Johnson, J. G. 1974. Extinction of perched faunas. *Geology* 2:479–482.

Jokat, W., G. Uenselmann-Neben, Y. Kristoffersen, and R. M. Rasmussen. 1992. Lomonosov Ridge—a double-sided continental margin. *Geology* 20:887–890.

Kapoor, H. M. 1992. Permo-Triassic boundary of the Indian subcontinent and its intercontinental correlation. In W. C. Sweet, Z. Y. Yang, J. M. Dickins, and H. F. Yin, eds., *Permo-Triassic Events in the Eastern Tethys*, pp. 21–36. Cambridge: Cambridge University Press.

Kapoor, H. M., and T. Tokuoka. 1985. Sedimentary facies of the Permian and Triassic of the Himalayas. In K. Nakazawa and J. M. Dickins, eds., *The Tethys: Her Paleogeography and Paleobiogeography from Paleozoic to Mesozoic*, pp. 23–58. Tokyo: Tokai University Press.

Kasting, J. F. 1992. Paradox lost and paradox found. *Nature* 355:676–677.

298

Kauffman, E. 1992. Personal communication.

Keith, M. L. 1982. Violent volcanism, stagnant oceans and some inferences regarding petroleum, strata-bound ores and mass extinctions. *Geochimica et Cosmochimica Acta* 46:2621–2637.

Kennedy, W. J. 1977. Ammonite evolution. In A. Hallam, ed., *Patterns of Evolution*, pp. 251–304. Amsterdam: Elsevier.

Kent, R. W., M. Storey, and A. D. Saunders. 1992. Large igneous provinces: sites of plume impact or plume incubation. *Geology* 20:891–894.

Keyser, A. W., and R. M. H. Smith. 1977. Vertebrate biozonation of the Beaufort Group with special reference to the Western Karoo Basin. *Annals of the South African Geological Survey* 12:1–36.

Kier, P. M. 1973. The echinoderms and Permian-Triassic time. In A. Logan and L. V. Hills, eds., *The Permian and Triassic Systems and Their Mutual Boundary*. Memoir 2, pp. 622–629. Calgary: Canadian Society of Petroleum Geologists.

Kier, P. M. 1984. Echinoids from the Triassic (St. Cassian) of Italy, their lantern supports and a revised phylogeny of Triassic echinoids. *Smithsonian Contributions to Paleobiology* 56:1–41.

King, G. M. 1990a. Dicynodonts and the end Permian event. *Palaeontologia Africa* 27:31–39.

King, G. M. 1990b. Life and Death in the Permo-Triassic: the fortunes of the Dicynodont mammal-like reptiles. *Sidney Haughtton Memorial Lecture, South African Museum.*

King, G. M. 1991. Terrestrial tetrapods and the end Permian event: a comparison of analyses. *Historical Biology* 5:239–255.

Kitchell, J. A., and T. R. Carr. 1985. Nonequilibrium model of diversification: faunal turnover dynamics. In J. W. Valentine, ed., *Phanerozoic Diversity Patterns*, pp. 277–309. Princeton: Princeton University Press.

Kitching, J. W. 1977. Distribution of the Karoo vertebrate fauna. *Memoirs of the Bernard Price Institute for Paleontological Research* 1:1–131.

Knoll, A. H. 1984. Patterns of extinction in the fossil record of vascular plants. In M. H. Nitecki, ed., *Extinctions*, pp. 23–68. Chicago: University of Chicago Press.

Koch, C. F. 1987. Prediction of sample size effects on the measured temporal and geographic distribution patterns of species. *Paleobiology* 13:100–107.

Koch, C. F., and J. P. Morgan. 1988. On the expected distribution of species' ranges. *Paleobiology* 14:126–138.

Kotljar, G. V. 1991. Permian-Triassic boundary in Tethys and Pacific belt and its correlation. *Proceedings of Shallow Tethys* 3:387–391.

Kozur, H. 1989. The Permian-Triassic boundary in marine and continental sediments. *Zbl.Geol. Palaont.* Teil I 1988:1245–1277.

Kramm, U., and K. H. Wedepohl. 1991. The isotopic composition of strontium and sulfur in seawater of Late Permian (Zechstein) age. *Chemical Geology* 90:253–262.

REFERENCES

Kummel, B. 1972. The Lower Triassic (Scythian) ammonoid *Otoceras. Bulletin of the Museum of Comparative Zoology.* 143:365–417.

Kummel, B. 1973a. Aspects of the Lower Triassic (Scythian) Stage. In A. Logan, and L. V. Hills, eds., *The Permian and Triassic Systems and their Mutual Boundary,* pp. 557–571. Calgary: Canadian Society of Petroleum Geologists.

Kummel, B. 1973b. Lower Triassic (Scythian) molluscs. In A. Hallam, ed., *Atlas of Palaeobiogeography,* pp. 225–233. Amsterdam: Elsevier.

Kummel, B., and C. Teichert. 1966. Relations between the Permian and Triassic formations in the Salt and Trans-Indus ranges, West Pakistan. *Neues Jahrbuch fur Geologie und Paläontologie* 125:297–333.

Kummel, B., and C. Teichert, eds. 1970a. Stratigraphic boundary problems: Permian and Triassic of West Pakistan. *University of Kansas Department of Geology, Special Publication* 4:474.

Kummel, B., and C. Teichert. 1970b. Stratigraphy and paleontology of the Permian-Triassic boundary Beds, Salt Range and Trans-Indus Ranges, West Pakistan. *University of Kansas, Department of Geology Special Publication* 4:1–110.

Kummel, B. and C. Teichert. 1973. The Permian-Triassic boundary in Central Tethys. In A. Logan and L. V. Hills, eds., *The Permian and Triassic Systems and Their Mutual Boundary.* Memoir 2, pp. 17–34. Calgary: Canadian Society of Petroleum Geologists.

Kump, L. R. 1991. Interpreting carbon-isotope excursions: Strangelove oceans. *Geology* 19:299–302.

Kutzbach, J. E., and Gallimore, R. G. 1989. Pangean climates: megamonsoons of the megacontinent. *Journal of Geophysical Research* 94:3341–3357.

Kutzbach, J. E., P. J. Guetter, W. F. Ruddiman, and W. L. Prell. 1989. Sensitivity of climate to Late Cenozoic uplift in southern Asia and the American West: numerical experiments. *Journal of Geophysical Research* 94:18,393–18,407.

Kvenvolden, K. 1988. Methane hydrate—a major reservoir of carbon in the shallow geosphere. *Chemical Geology* 71:41–51.

Labandeira, C. 1992. Personal communication.

Labandeira, C. C., and B. S. Beall. 1990. Arthropod terrestriality. In D. G. Mikulic, ed., *Arthropod Paleobiology, Notes from a Short Course,* pp. 214–256. Knoxville, Tenn.:Paleontological Society.

Lantzy, R. J., M. F. Dacey, and F. T. MacKenzie. 1977. Catastrophe theory: application to the Permian mass extinction. *Geology* 5:724–728.

Larson, R. L. 1991a. Geological consequences of superplumes. *Geology* 19:963–966.

Larson, R. L. 1991b. Latest pulse of Earth: evidence for a mid-Cretaceous superplume. *Geology* 19:547–550.

Larson, R. L., and P. Olson. 1991. Mantle plumes control magnetic reversal frequency. *Earth and Planetary Science Letters* 107:437–447.

Larson, R. L., and W. C. Pitman. 1972. Worldwide correlation of Mesozoic magnetic anomalies, and its implications. *Geological Society of America Bulletin* 83:3645–3662.

Lashof, D. A. 1991. Gaia on the brink: biogeochemical feedback processes in global warming. In S. S. Schneider and P. A. Boston, eds., *Scientists on Gaia*, pp. 393–404. Cambridge, Mass.: MIT Press.

LeFort, J. P., and R. Van der Voo. 1981. A kinematic model for the collision and complete suturing between Gondwanaland and Laurasia in the Carboniferous. *Journal of Geology* 89:537–550.

Li, Z. S., et al. 1989. Study on the Permian-Triassic biostratigraphy and event stratigraphy of northern Sichuan and southern Shanaxi. *Geological Memoirs*, Series 2, No. 9.

Li, Z. S., L. P. Zhan, J. X. Yao, and Y. Q. Zhou. 1991. On the Permian-Triassic events in South China—probe into the end Permian abrupt extinction and its possible causes. *Proceedings of Shallow Tethys* 3:371–382.

Li, Z. S., L. P. Zhan, X. F. Zhu, L. C. Xie, G. F. Liu, J. H. Zhang, R. G. Jin, and H. Q. Huang. 1986. Mass extinctions and geological events between Paleozoic and Mesozoic eras. *Acta Geologica Sinica* 60:1–17.

Liao, Z. T. 1980. Brachiopod assemblages from the Upper Permian and Permian-Triassic boundary beds, South China. *Canadian Journal of Earth Sciences* 17:289–295.

Lin, J. L., M. Fuller, and W. Y. Zhang. 1985. Preliminary Phanerozoic polar wander paths for the North and South China blocks. *Nature* 313:444–449.

Lipps, J. H. 1970. Plankton evolution. *Evolution* 24:1–22.

Logan, A., and Hills, L. V., eds. 1973. *The Permian and Triassic Systems and Their Mutual Boundary*. Memoir 2. Calgary: Canadian Society of Petroleum Geologists.

Loper, D. E. 1992. On the correlation between mantle plume flux and the frequency of reversals of the geomagnetic field. *Geophysical Research Letters* 19:25–28.

Loper, D. E., and K. McCartney. 1990. On impacts as a cause of geomagnetic field reversals or flood basalts. In V. L. Sharpton, and P. D. Ward, eds., *Global Catastrophes in Earth History*, pp.19–26. Boulder, Colo.: Geological Society of America.

Loper, D. E., K. McCartney, and G. Buzyana. 1988. A model of correlated episodicity in magnetic-field reversals, climate and mass extinctions. *Journal of Geology* 96:1–15.

Lottes, A. L., and D. B. Rowley. 1990. Reconstruction fo the Laurasian and Gondwanan segments of Permian Pangaea. In W. S. McKerrow and C. R. Scotese, eds., *Palaeozoic Palaeogeography and Biogeography*, pp. 383–395. Geological Society Memoir No. 12. London: Geological Society.

Luhr, J. F. 1991. Volcanic shade causes cooling. *Nature* 354:104–105.

MacArthur, R. H., and E. O. Wilson. 1967. *The Theory of Island Biogeography*. Princeton: Princeton University Press.

REFERENCES

McCammon, H. M. 1969. The food of articulate brachiopods. *Journal of Paleontology* 43:976–985.

McCartney, K., A. R. Huffman, and M. Tredoux. 1990. A paradigm for endogenous causation of mass extinctions. In V. L. Sharpton and P. D. Ward, eds., *Global Catastrophes in Earth History*, pp. 125–138. Boulder, Colo.: Geological Society of America.

MacKenzie, F. T., and J. D. Pigott. 1981. Tectonic controls of Phanerozoic sedimentary cycling. *Journal of the Geological Society of London* 138:183–196.

McKerrow, W. S., and C. R. Scotese, eds. 1990. *Palaeozoic Palaeogeography and Biogeography*. Geological Society Memoir No. 12. London: Geological Society.

McLean, D. M. 1985. Deccan traps mantle degassing in the terminal Cretaceous marine extinctions. *Cretaceous Research* 6:235–259.

Magaritz, M. 1989. ^{13}C minima follow extinction events: a clue to faunal radiation. *Geology* 17:337–340.

Magaritz, M. 1992. Personal communication.

Magaritz, M., R. Y. Anderson, W. T. Holser, E. S. Saltzman, and J. Garber. 1983. Isotopic shifts in the Late Permian of the Delaware Basin, Texas, precisely timed by varved sediments. *Earth and Planetary Science Letters* 66:111–124.

Magaritz, M., R. Bär, A. Baud, and W. T. Holser. 1988. The carbon-isotope shift at the Permian/Triassic boundary in the southern Alps is gradual. *Nature* 331:337–339.

Magaritz, M., R. V. Krishnamurthy, and W. T. Holser. Forthcoming. Parallel trends in organic and inorganic isotopes across the Permian/Triassic boundary. *American Journal of Science*.

Makarenko, G. F. 1976. The epoch of Triassic trap magmatism in Siberia. *International Geology Review* 19:1089–1100.

Małkowski, K., M. Gruszczyński, A. Hoffman, and S. Halas. 1989. Oceanic stable iostope composition and a scenario for the Permo-Triassic crisis. *Historical Biology* 2:289–309.

Małkowski, K., M. Gruszczyński, and A. Hoffman. 1991. A facies geological test of stable isotope interpretation of the Upper Permian depositional environment in West Spitsbergen. *Terra Nova* 3:631–637.

Mamay, S., and S. Wing. 1992. Personal communication.

Manabe, S., and A. J. Broccoli. 1990. Mountains and arid climates of middle latitudes. *Science* 247:192–194.

Marcoux, J., and A. Baud. 1988. The Permo-Triassic boundary in the Antalya Nappes (western Taurides, Turkey). *Memorie della Societa Geologicà Italiana* 34:243–252.

Marshall, C. R. 1990. Confidence intervals on stratigraphic ranges. *Paleobiology* 16:1–10.

Maxwell, W. D. 1989. The end Permian mass extinction. In S. K. Donovan, ed., *Mass Extinctions: Processes and Evidence*, pp. 152–173. London: Belhaven Press.

Maxwell, W. D. 1992. Permian and Early Triassic extinction of non-marine tetrapods. *Palaeontology* 35:571–584.

Maxwell, W. D., and M. J. Benton. 1990. Historical tests of the absolute completeness of the fossil record of tetrapods. *Paleobiology* 16:322–335.

Meldahl, K. H. 1990. Sampling, species abundance, and the stratigraphic signiture of mass extinction: a test using Holocene tidal flat molluscs. *Geology* 18:890–893.

Menning, M. 1990. A new scheme for the Permian and Triassic succession of central Europe. *Permophiles* 16:14.

Menning, M. 1991. Numerical time scale for the Permian. *Permophiles* 20:2–3.

Metcalfe, I. 1984. The Permian-Triassic boundary in northwest Malaya. *Warti Geologi* 10:139–145.

Metcalfe, I. 1988. Origin and assembly of south-east Asian continental terranes. In M. G. Audley-Charles and A. Hallam, eds., *Gondwana and Tethys*, pp. 101–118. London: Geological Society of London.

Meyen, S. F. 1973. The Permian-Triassic boundary and its relation to the Paleophyte-Mesophyte floral boundary. In A. Logan and L. V. Hills, eds., *The Permian and Triassic Systems and Their Mutual Boundary*, Memoir 2, pp. 662–667. Calgary: Canadian Society of Petroleum Geologists.

Meyen, S. V. 1982. The Carboniferous and Permian floras of Angaraland (a synthesis). *Biological Memoirs* 7:1–110.

Middlemiss, C. S. 1909. Gondwana and related sedimentary systems of Kashmir. *Records of the Geological Society of India* 37:286–327.

Mikolajewicz, U., B. D. Santer, and E. Maier-Reimer. 1990. Ocean response to greenhouse warming. *Nature* 345:589–593.

Miller, A. I. 1988. Spatio-temporal transitions in Paleozoic Bivalvia: an analysis of North American fossil assemblages. *Historical Biology* 1:251–274.

Miller, A. I. 1989. Spatio-temporal transitions in Paleozoic Bivalvia: a field comparison of Upper Ordovician and Upper Paleozoic bivalve-dominated fossil assemblages. *Historical Biology* 2:227–260.

Milner, A. R. 1990. The radiation of temnospondyl amphibians. In P. D. Taylor and G. P. Larwood, eds., *Major Evolutionary Radiations*, pp. 321–349. Oxford: Clarendon Press.

Mojsisovics, E. 1879. Vorläufige kurze Übersicht der Ammoniten-Gattungen der mediterranen und juvavischen Trias. *Verhandl. Geol. Reichsanst. Wien.* 1879:133–143.

Mojsisovics, E., W. von Waagen, and K. Diener. 1895. Entwurf einer Gliederung der pelagischen Sedimente des Trias-Systems. *Sitzungberichte Akademie Wissenshaften Wien* 104:1271–1302.

Molnar, P., and P. England. 1990. Late Cenozoic uplift of mountain ranges and global climate change: chicken or egg. *Nature* 346:29–34.

REFERENCES

Morel, P., and E. Irving. 1981. Paleomagnetism and the evolution of Pangea. *Journal of Geophysical Research* 86:1858–1872.

Morgan, L. A., and W. Bonnichsen. 1989. Heise volcanic field. In C. E. Chapin and J. Zidek, *Field Excursions to Volcanic Terranes in the Western United States, Vol. II: Cascades and the Intermountain West.* Memoir 47, pp. 153–160. Soccorro, N.M.: New Mexico Bureau of Mines and Mineral Resources.

Morgan, L. A., D. J. Doherty, and W. P. Leeman. 1984. Ignimbrites of the eastern Snake River Plain: evidence for major-caldera forming earuptions. *Journal of Geophysical Research* 89:8665–8678.

Morgan, W. J. 1981. Hot spot tracks and the opening of the Atlantic and Indian Oceans. In C. Emiliani, ed. *The Oceanic Lithosphere. The Sea*, vol. 7, pp. 443–487. New York: Wiley.

Murchison, R. I. 1841. Letter to M. Fischer de Waldheim. *Philisophical Magazine* 19:418–422.

Murchison, R. I., E. De Verneuil, and A. von Keyserling. 1845. *The Geology of Russia in Europe and the Ural Mountains.* London: John Murray.

Nakazawa, K. 1981. Analysis of Late Permian and Early Triassic Faunas of Kashmir. *Palaeontologia Indica, NS* 46:191–204.

Nakazawa, K., Y. Bando, and T. Matsuda. 1980. The *Otoceras woodwardi* Zone and the time-gap at the Permian-Triassic boundary in East Asia. *Geology and Paleontology of Southeast Asia* 21:75–90.

Nakazawa, K., and J. M. Dickins, eds. 1985. *The Tethys: Her Paleogeography and Palaeobiogeography from Paleozoic to Mesozoic.* Tokyo: Tokai University Press.

Nakazawa, K., K. I. Ishii, M. Kato, Y. Okimura, K. Kakamura, and D. Haralombous. 1975. Upper Permian fossils from island of Salamis, Greece. *Memoirs of the Faculty of Science, Kyoto University, Series of Geology and Mineralogy* 41(2):21–44.

Nakazawa, K., and H. M. Kapoor, eds. 1981. The Upper Permian and Lower Triassic faunas of Kashmir. *Palaeontologia Indica, NS* 46:1–203.

Nakazawa, K., H. M. Kapoor, K. I. Ishii, Y. Bando, Y. Okimura, T. Tokuoka, M. Murata, K. Nakamura, Y. Nogami, S. Sakagami, and D. Shimizu. 1974. The Upper Permian and the Lower Triassic in Kashmir, India. *Memoirs of the Faculty of Science, Kyoto University, Series of Geology and Mineralogy* 42:1–106.

Nakazawa, K., and B. Runnegar. 1973. The Permian-Triassic boundary: a crisis for bivalves? In A. Logan and L. V. Hills, eds., *The Permian and Triassic System and Their Mutual Boundary.* Memoir 2, pp. 608–621. Calgary: Canadian Society of Petroleum Geologists.

Nance, R. D., T. R. Worsley, and J. B. Moody. 1986. Post-Archean biogeochemical cycles and long-term episodicity in tectonic processes. *Geology* 14:514–518.

Nance, R. D., T. R. Worsley, and J. B. Moody. 1988. The supercontinent cycle. *Scientific American* 00:72–79.

Newell, N. D. 1956. Catastrophism and the fossil record. *Evolution* 10:97–101.

Newell, N. D. 1962. Paleontological gaps and geochronology. *Journal of Paleontology* 36:592–610.

Newell, N. D. 1967a. Paraconformities. In C. Teichert and E. L. Yochelson, eds. *Raymond C. Moore Commemorative Volume. Essays in Palaeontology and Stratigraphy*, pp. 349–367 Lawrence: University of Kansas Press.

Newell, N. D. 1967b. Revolutions in the history of life. *Geological Society of America Special Paper* 89:63–91.

Newell, N. D. 1973. The very last moment of the Paleozoic era. In A. Logan and L. V. Hills, eds., *The Permian and Triassic System and Their Mutual Boundary*. Memoir 2, pp. 1–10. Calgary: Canadian Society of Petroleum Geologists.

Newell, N. D. 1978. The search for a Paleozoic-Mesozoic boundary stratotype. *Schriftenreihe Erdwiss. Komm. Oster. Akad. Wiss.* 4:9–19.

Newell, N. D. 1982. Mass extinctions—illusions or realities?. In L. T. Silver, and P. H. Schultz, eds., *Geological Implications of Impacts of Large Asteroids and Comets on the Earth*, pp. 257–263. Geological Society of America Special Paper 190. Boulder, Colo.: Geological Society of America.

Newell, N. D. 1988. The Paleozoic/Mesozoic erathem boundary. *Memorie della Società Geologica Italiana* 34:303–311.

Nie, S. Y., D. B. Rowley, and A. M. Ziegler. 1990. Constraints on the locations of Asian microcontinents in Palaeo-Tethys during the Late Paleozoic. In W. S. McKerrow, and C. R. Scotese, eds., *Palaeozoic Palaeogeography and Biogeography*, pp. 397–409. London: Geological Society Memoir.

Niklas, K. J., B. H. Tiffney, and A. H. Knoll. 1985. Patterns in vascular land plant diversification: an analysis at the species level. In J. W. Valentine, ed., *Phanerozoic Diversity Patterns*, pp. 97–128. Princeton: Princeton University Press.

Nisbet, E. G. 1990. The end of the ice age. *Canadian Journal of Earth Sciences* 27:148–157.

Noè, S. 1987. Facies and paleogeography of the Marine Upper Permian and of the Permian-Triassic boundary in the southern Alps (Bellerophon Formation, Tesero Horizon). *Facies* 16:89–142.

Oberhansli, H., K. J. Hsu, S. Piasecki, and H. Weissert. 1989. Permian-Triassic carbon-isotope anomaly in Greenland and in the southern Alps. *Historical Biology* 2:37–49.

Oliver, W. A., Jr. 1980. The relationship of the scleractinian corals to the rugose corals. *Paleobiology* 6:146–160.

Olson, E. C. 1982. Extinctions of Permian and Triassic nonmarine vertebrates. In L. T. Silver, and P. H. Schultz, eds., *Geological Implications of Impacts of Large Asteroids and Comets on the Earth*, pp. 501–511. Geological

Society of America Special Paper 190. Boulder, Colo.: Geological Society of America.

Olson, E. C. 1989. Problems of Permo-Triassic terrestrial vertebrate extinctions. *Historical Biology* 2:17–35.

Olson, P., P. G. Silver, and R. W. Carlson. 1990. The large-scale structure of convection in the Earth's mantle. *Nature* 344:209–215.

Orth, C. J. 1989. Geochemistry of the bio-event horizons. In S. K. Donovan, ed., *Mass Extinctions: Processes and Evidence*, pp. 37–72. London: Belkhaven Press.

Orth, C. J., M. Attrep, Jr., and L. R. Quintana. 1990. Iridium abundance patterns across bio-event horizons in the fossil record. In V. L. Sharpton and P. D. Ward, eds., *Global Catastrophes in Earth History*, pp. 45–60. Geological Society of America Special Paper 247. Boulder, Colo.: Geological Society of America.

Owen, H. G. 1983. Ocean-floor spreading evidence of global expansion. In S. E. Carey, ed., *Expanding Earth Symposium*, pp. 31–58. Sydney: University of Tasmania.

Padian, K., and W. A. Clemens. 1985. Terrestrial vertebrate diversity: episodes and insights. In J. W. Valentine, ed., *Phanerozoic Diversity Patterns*, pp. 41–69. Princeton: Princeton University Press.

Pakistani-Japanese Research Group. 1985. Permian and Triassic Systems in the Salt Range and Surghar Range, Pakistan. In K. Nakazawa and J. M. Dickins, eds. *The Tethys: Her Paleogeography and Paleobiogeography from Paleozoic to Mesozoic*, pp. 219–311. Toyko: Tokai University Press.

Palais, J. M., and H. Sigurdsson. 1989. Petrologic evidence of volatile emissions from major historic and pre-historic volcanic eruptions. In A. Berger, R. E. Dickinson, and J. W. Kidson, eds., *Understanding Climate Change*, pp.31–53. Washington, D. C.: American Geophysical Union.

Pantić-Prodanović, S. 1989/90. Micropaleontologic and biostratigraphic of Upper Permian and Lower Triassic sediments in northwestern Serbia, Yugoslavia. *Glasnik Prirodnjačkog Muzeja U Beogradu A* 44/45:172–175.

Parrington, F. R. 1948. Labyrinthodonts from South Africa. *Proceedings of the Zoological Society of London* 118:426–445.

Parrish, J. M., J. T. Parrish, and A. M. Ziegler. 1986. Permian-Triassic paleogeography and paleoclimatology and implications for Therapsid distribution. In N. Hotton, P. D. MacLean, J. J. Roth, and E. C. Roth, eds., *The Ecology and Biology of Mammal-Like Reptiles*, pp. 109–131. Washington, D.C.: Smithsonian Institution Press.

Parrish, J. T. 1982. Upwelling and petroleum source beds, with reference to Paleozoic. *American Association of Petroleum Geologists Bulletin* 66:750–774.

Parrish, J. T., and R. L. Curtis. 1982. Atmospheric circulation, upwelling and organic-rich rocks in the Mesozoic and Cenozoic. *Palaeogeography, Palaeoclimatology, Palaeoecology* 40:31–66.

Patterson, C., and A. B. Smith. 1989. Periodicity in extinctions: the role of systematics. *Ecology* 70:802–811.

Paul, C. R. C. 1988. Extinction and survival in the echinoderms. In G. P. Larwood, ed. *Extinction and Survival in the Fossil Record*, pp. 155–170. Oxford: Clarendon Press.

Paulay, G. 1990. Effects of late Cenozoic sea-level fluctuations on the bivalve faunas of tropical oceanic islands. *Paleobiology* 16:415–434.

Paull, C. K., W. Ussler, III, and W. P. Dillon. 1991. Is the extent of glaciation limited by marine gas-hydrates? *Geophysical Research Letters* 18:432–434.

Pešič, L., A. Ramovš, J. Sremac, S. Pantić-Prodanović, I. Filipovič, S. Kovács, and P. Pelikán. 1988. Upper Permian deposits in the Jadar region and their position within the western Paleotethys. *Memorie della Società Geologica Italiana* 34:211–220.

Phillips, J. 1840a. Organic remains. *Penny Cyclopedia* 16:487–491.

Phillips, J. 1840b. Paleozoic series. *Penny Cyclopedia* 17:153–154.

Phillips, J. 1841. *Figures and Descriptions of the Palaeozoic Fossils of Cornwall, Devon and West Somerset; Observed in the Course of the Ordnance Survey of that District*. London: Longman.

Phillips, J. 1860. *Life on the Earth. Its Origin and Succession*. Cambridge: Macmillan.

Piasecki, S., and C. Marcussen. 1986. Oil geological studies in the central East Greenland. *Rapp. Grønlands geolgiske Undersøgelse* 130:95–102.

Pinto, J. P., R. P. Turco, and O. B. Toon. 1989. Self-limiting physical and chemical effects in volcanic eruption clouds. *Journal of Geophysical Research* 94:11,165–11,174.

Pitrat, C. W. 1970. Phytoplankton and the late Paleozoic wave of extinction. *Palaeogeography, Palaeoclimatology, Palaeoecology* 8:49–55.

Pitrat, C. W. 1973. Vertebrates and the Permo-Triassic extinction. *Palaeogeography, Palaeoclimatology, Palaeoecology* 14:249–264.

Ponder, W. F., and A. Waren. 1988. Classification of the Caenogastropoda and heterostropha—a list of the family-group names and higher taxa. *Malacological Review, Supplement* 4:288–326.

Popp, B. N., F. A. Podosek, J. C. Brannon, T. F. Anderson, and J. Pier. 1986. $^{87}Sr/^{86}Sr$ ratios in Permo-Carboniferous sea water from the analyses of well-preserved brachiopod shells. *Geochimica et Cosmochimica Acta* 50:1321–1328.

Posenato, R. 1991. Endemic to cosmopolitan brachiopods across the P/Tr boundary in the southern Alps (Italy). *Saito Ho-on Kai Special Publication, (Proceedings of Shallow Tethys 3)* 3:125–139.

Quinn, J. F. 1987. On the statistical detection of cycles in extinction in the marine fossil record. *Paleobiology* 13:465–478.

Rampino, M. R., and S. Self. 1982. Historic eruptions of Tampora (1815), Krakatau (1883) and Agung (1963), their stratospheric aerosols and climatic impact. *Quarternary Research* 18:127–143.

REFERENCES

Rampino, M. R., and S. Self. 1991. Volcanic winter and accelerated glaciation following the Toba super-eruption. *Nature* 359:50–52.

Rampino, M. R., S. Self, and R. B. Strothers. 1988. Volcanic winters. *Annual Review of Earth and Planetary Sciences* 16:73–99.

Rampino, M. R., and R. B. Strothers. 1988. Flood basalt volcanism during the past 250 million years. *Science* 241:663–668.

Raup, D. M. 1976. Species diversity in the Phanerozoic: an interpretation. *Paleobiology* 2:289–297.

Raup, D. M. 1978. Cohort analysis of generic survivorship. *Paleobiology* 4:1–15.

Raup, D. M. 1979. Size of the Permo-Triassic bottleneck and its evolutionary implications. *Science* 206:217–218.

Raup, D. M. 1982. Biogeographic extinction: a feasibility test. In L. T. Silver and P. H. Schultz, eds., *Geological Implications of Impacts of Large Asteroids and Comets on the Earth*, pp. 272–282. Special Paper 190. Boulder, Colo.: Geological Society of America.

Raup, D. M. 1985. Magnetic reversals and mass extinctions. *Nature* 314:341–343.

Raup, D. M. 1987. Mass extinction: a commentary. *Palaeontology* 30:1–13.

Raup, D. M. 1991. *Extinction: Bad Genes or Bad Luck?* New York: Norton.

Raup, D. M. 1992. Large-body impact and extinction in the Phanerozoic. *Paleobiology* 18:80–88.

Raup, D. M., and G. E. Boyajian. 1988. Patterns of generic extinction in the fossil record. *Paleobiology* 14:109–125.

Raup, D. M., and J. J. Sepkoski, Jr. 1982. Mass extinction in the marine fossil record. *Science* 215:1501–1503.

Raup, D. M., and J. J. Sepkoski, Jr. 1984. Periodicity of extinctions in the geologic past. *Proceedings of the National Academy of Sciences USA* 81:801–805.

Raup, D. M., and J. J. Sepkoski, Jr. 1986a. Periodic extinctions of families and genera. *Science* 231:833–836.

Raup, D. M., and J. J. Sepkoski, Jr. 1986b. Testing for periodicity of extinction. *Science* 241:94–96.

Raymo, M. E. 1991. Geochemical evidence supporting T.C. Chamberlin's theory of glaciation. *Geology* 19:344–347.

Raymo, M. E., and W. F. Ruddiman. 1992. Tectonic forcing of late Cenozoic climate. *Nature* 359:117–122.

Raymo, M. E., W. F. Ruddiman, and P. N. Froelich. 1988. Influence of late Cenozoic mountain building on ocean geochemical cycles. *Geology* 16:649–653.

Reinhardt, J. W. 1988a. End-Paleozoic regression and carbonate to siliciclastic transition: a Permian-Triassic boundary on the Eastern Tethyan Yangzi Platform (Sichuan, China). *Zbl. Geol. Palaont* 1988:861–870.

Reinhardt, J. W. 1988b. Uppermost Permian reefs and Permo-Triassic sedimentary facies from the southeastern margin of Sichuan Basin, China. *Facies* 18:231–288.

Renne, P. R., and A. R. Basu. 1991. Rapid eruption of the Siberian Traps flood basalts at the Permo-Triassic boundary. *Science* 253:176–179.

Rhodes, F. H. T. 1967. Permo-Triassic extinction. In W. B. Harland, C. H. Holland, M. R. House, N. F. Hughes, A. B. Reynolds, M. J. S. Rudwick, G. E. Satterthwaite, L. B. H. Tarlo, and E. C. Wiley, eds. *The Fossil Record*, pp. 55–76. London: Geological Society.

Rhodes, M. C., and C. W. Thayer. 1991. Mass extinctions:ecological selectivity and primary production. *Geology* 19:877–880.

Richards, M. A., R. A. Duncan, and V. E. Courtillot. 1989. Flood basalts and hot-spot tracks: plume heads and tails. *Science* 246:103–109.

Robinson, P. L. 1973. Palaeoclimatology and continental drift. In D. H. Darlington and S. K. Runcorn, eds., *Implications of Continental Drift to the Earth Sciences*, pp. 451–476. London: Academic Press.

Rohdendorf, B.B., and A.P. Rasnitsyn, eds. 1980. Historical development of the Class Insecta. *Trudy Palaeontological Institute, Moscow* 175:1-250 (in Russian)

Romer, A.S. 1966. *Vertebrate Paleontology*, 3d ed. Chicago: University of Chicago Press.

Ross, C. A., and J. R. P. Ross. 1985. Late Paleozoic depositional sequences are synchronous and worldwide. *Geology* 13:194–197.

Ross, C. A., and J. R. P. Ross. 1987. Late Paleozoic sea levels and depositional seqeunces. In C. A. Ross and D. Haman, eds. *Timing and Depositional History of Eustatic Sequences: Constraints on Seismic Stratigraphy*, pp. 137–149. Special Publication No. 24. Washington, D.C.: Cushman Foundation for Foraminiferal Research.

Ross, J. R. P. 1978. Biogeography of Permian ectoproct bryozoa. *Palaeontology* 21:341–356.

Rostovtsev, K. O., and N. R. Azaryan. 1973. The Permian-Triassic boundary in Transcaucasia. In A. Logan and L. V. Hills, eds. *The Permian and Triassic System and Their Mutual Boundary*. Memoir 2, pp. 89–99. Calgary: Canadian Society of Petroleum Geologists.

Ruddiman, W. F., and J. E. Kutzbach. 1989. Forcing of late Cenozoic Northern Hemisphere climate by plateau uplift in southern Asia and the American West. *Journal of Geophysical Research* 94:18,409–18427.

Ruddiman, W. F., and J. E. Kutzbach. 1991. Plateau uplift and climatic change. *Scientific American* 264(3):66–75.

Ruddiman, W. F., W. L. Prell, and M. E. Raymo. 1989. Late Cenozoic uplift in southern Asia and the American West: rationale for general circulation modeling experiments. *Journal of Geophysical Research* 94:18,379–18,391.

Rudwick, M. J. S. 1985a. *The Great Devonian Controversy*. Chicago: University of Chicago Press.

REFERENCES

Rudwick, M. J. S. 1985b. *The Meaning of Fossils*. Chicago: University of Chicago Press.

Rui, L., J. W. He, C. Z. Chen, and Y. G. Wang. 1988. Discovery of fossil animals from the basal clay of Permian-Triassic boundary in the Meishan area of Changzing, Zhejiang and its significance. *Journal of Stratigraphy* 12(1):48–53.

Rutland, C., H. W. Smedes, R. I. Tilling, and W. R. Greenwood. 1989. Volcanism and plutonism at shallow crustal levels: the Elkhorn Mountian volcanics and the Boulder Batholith, southwestern Montana. In C. E. Chapin and J. Zidek, *Field Excursions to Volcanic Terranes in the Western United States, Vol. II: Cascades and the Intermountian West*, Memoir 47, pp. 264–276. Soccorro, N.M.: New Mexico Bureau of Mines and Mineral Resources.

Ruzhentsev, V. E., and T. G. Sarycheva. 1965. Razvitie i smena morskikh organizmov na rubezhe Paleozoya i Mezozoya [The development and change of marine organisms at the Paleozoic-Mesozoic boundary]. *Trudy Paleontol. Instiute Akad. Nauk SSSR* 180:1–431. Translated by D. A. Brown, Geology Department, Australian National Museum. Publication No. 117, 1968.

Sadler, P. 1981. Sediment accumulation rates and the completeness of stratigraphic sections. *Journal of Geology* 89:569–84.

Sakagami, S. 1985. Paleogeographic distribution of Permian and Triassic Ectoprocta (Bryozoa). In K. Nakazawa and J. M. Dickins, eds., *The Tethys: Her Paleogeography and Paleobiogeography from Paleozoic to Mesozoic*, pp. 171–184. Tokyo: Tokai University Press.

Schaeffer, B. 1973. Fishes and the Permian-Triassic boundary. In A. Logan and L. V. Hills, eds. *The Permian and Triassic System and Their Mutual Boundary*. Memoir 2, pp. 493–497. Calgary: Canadian Society of Petroleum Geologists.

Schafer, P., and E. Fois-Erickson. 1986. Triassic bryozoa and the evolutionary crisis of Paleozoic Stenolaemata. In O. Walliser, ed., *Global Bio-Events*, pp. 251–255. Berlin: Springer-Verlag.

Schenck, H. G., T. S. Childs, R. W. Crume, K. L. Edwards, S. Holliday, J. G. Marks, S. W. Muller, W. A. Newton, and F. Paguirgan. 1941. Stratigraphic nomenclature. *Bulletin of the American Association of Petroleum Geologists* 25:2195–2202.

Schindewolf, O. H. 1954. Über die Faunenwende vom Palaozoikum zum Mesozoikum. *Zeitschrift Deutschen Geologischen Gesellschaft* 105:153–183.

Schindewolf, O. H. 1963. Neokatastrophismus? *Zeitschrift der Deutschen Geologischen Gesellschaft* 114:430–445.

Schopf, T. J. M. 1974. Permo-Triassic extinctions: relation to sea-floor spreading. *Journal of Geology* 82:129–143.

Schopf, T. J. M. 1979. The role of biogeographic provinces in regulating marine faunal diversity through geologic time. In J. Gray and A. J. Boucot, eds.

REFERENCES

Historical Biogeography, Plate Tectonics and the Changing Environ-ment, pp. 449–457. Corvallis: Oregon State University Press.

Schubert, J. K., and D. J. Bottjer. 1992. Early Triassic stromatolites as post-extinction disaster forms. *Geology* 20:883–886.

Schubert, J. K., D. J. Bottjer, and M. J. Simms. 1992. Paleobiology of the oldest known articulate crinoid. *Lethaia* 25:97–110.

Sclater, J. G., and J. Fancheteau. 1970. The implications of terrestrial heat flow observations on current tectonic and geochemical models of the crust and upper mantle of the earth. *Geophysical Journal of the Royal Astronomical Society* 20:509–542.

Scotese, C. R. 1984. Introduction: Paleozoic paleomagnetism and the assem-bly of Pangea. *American Geophysical Union Geodynamics Series* 12:1–10.

Scotese, C. R., and W. S. McKerrow. 1990. Revised world maps and introduc-tion. In W. S. McKerrow and C. R. Scotese, eds., *Palaeozoic Palaeogeog-raphy and Biogeography*, Memoir No. 12, pp. 1–21. London: Geological Society.

Scott, A. C., J. Stephenson, and W. G. Chaloner. 1992. Interaction and coe-volution of plants and arthropods during the Palaeozoic and Mesozoic. *Philosophical Transactions of the Royal Society, London. Series B* 335:129–165.

Scrutton, C. T. 1988. Patterns of extinction and survival in Paleozoic corals. In G. P. Larwood, ed., *Extinction and Survival in the Fossil Record*, pp. 65–88. Oxford: Clarendon Press.

Şengör, A. M. C. 1987. Tectonics of the Tethysides: orogenic collage devel-opment in a collisional setting. *Annual Review of Earth and Planetary Sciences* 15:213–244.

Şengör, A. M. C., D. Altiner, A. Cin, T. Ustaömer, and K. J. Hsü. 1988. Origin and assembly of the Tethyside orogenic collage at the expense of Gond-wana Land. In M. G. Audley-Charles and A. Hallam, *Gondwana and Tethys*, pp. 119–181. Geological Society Special Publication No. 37. Lon-don: Geological Society.

Secord, J. A. 1986. *Controversy in Victorian Geology: The Cambrian-Silurian Dispute*. Princeton: Princeton University Press.

Sepkoski, J. J., Jr. 1979. A kinetic model of Phanerozoic taxonomic diversity: II. Early Phanerozoic families and multiple equilibria. *Paleobiology* 5:222–251.

Sepkoski, J. J., Jr. 1981. A factor analytic description of the marine fossil rec-ord. *Paleobiology* 7:36–53.

Sepkoski, J. J., Jr. 1982. Mass extinctions in the Phanerozoic oceans: a review. In L. T. Silver and P. H. Schultz, eds., *Geological Implications of Impacts of Large Asteroids and Comets on the Earth*, pp. 283–290. Geological Society of America Special Paper 247. Boulder, Colo.: Geological Society of America.

Sepkoski, J. J., Jr. 1984. A kinetic model of Phanerozoic taxonomic diversity.

III. Post-Paleozoic families and mass extinctions. *Paleobiology* 10:246–267.

Sepkoski, J. J., Jr. 1986a. Global bioevents and the question of periodicity. In O. Walliser, ed., *Global Bio-Events*, pp. 47–61. Berlin: Springer-Verlag.

Sepkoski, J. J., Jr. 1986b. Phanerozoic overview of mass extinction. In D. M. Raup and D. Jablonski, eds., *Patterns and Processes in the History of Life*, pp. 277–295. Berlin: Springer-Verlag.

Sepkoski, J. J., Jr. 1989. Periodicity in extinction and the problem of catastrophism in the history of life. *Journal of the Geological Society of London* 146:7–19.

Sepkoski, J. J., Jr. 1990. The taxonomic structure of periodic extinction. In V. L. Sharpton and P. D. Ward, eds., *Global Catastrophes in Earth History*, pp. 33–44. Geological Society of America Special Paper 247. Boulder, Colo.: Geological Society of America.

Sepkoski, J. J., Jr., 1991. A model of onshore-offshore change in faunal diversity. *Paleobiology* 17:58–77.

Sepkoski, J. J., Jr. 1992a. A compendium of fossil marine animal families, 2d ed. *Milwaukee Public Museum Contributions in Biology and Geology* 83:155.

Sepkoski, J. J., Jr. 1992b. Personal communication.

Sepkoski, J. J., Jr. 1992c. Personal communication.

Sepkoski, J. J., Jr., R. K. Bambach, D. M. Raup, and J. W. Valentine. 1981. Phanerozoic marine diversity and the marine fossil record. *Nature* 293:435–437.

Sepkoski, J. J., Jr., and M. L. Hulver. 1985. An atlas of Phanerozoic clade diversity diagrams. In J. W. Valentine, ed., *Phanerozoic Diversity Patterns*, pp. 11–39. Princeton: Princeton University Press.

Sepkoski, J. J., Jr., and A. I. Miller. 1985. Evolutionary faunas and the distribution of Paleozoic marine communities in space and time. In J. W. Valentine, ed., *Phanerozoic Diversity Patterns*, pp. 153–189. Princeton: Princeton University Press.

Sepksoki, J. J., Jr., and D. M. Raup. 1986. Periodicity in marine extinction events. In. D. K. Elliott, ed., *Dynamics of Extinction*, pp. 3–36. New York: John Wiley.

Shear, W. A., and J. Kukalova-Peck. 1990. The ecology of Paleozoic terrestrial arthropods: the fossil evidence. *Canadian Journal of Zoology* 68:1807–1834.

Sheehan, P. M. 1985. Reefs are not so different—they follow the evolutionary pattern of level-bottom communities. *Geology* 13:46–49.

Sheng, J. Z., C. Z. Chen, Y. G. Wang, and R. Lin. 1987. New advances on the Permian and Triassic boundary of Jiangsu, Zhejiang and Anhui. In *Stratigraphy and Palaeontology of Systemic Boundaries in China, Permian-Triassic Boundary, 1*, pp. 1–19. Nanjing: Nanjing University Press.

Sheng, J. Z., C. Z. Chen, Y. G. Wang, L. Rui, Z. T. Liao, Y. Bando, K. I. Ishii, K. Nakazawa, and K. Nakamura. 1984. Permian-Triassic boundary in middle and Eastern Tethys. *Journal of the Faculty of Science of Hokkaido University, Ser. IV* 21:133–181.

Sheng, J. Z., L. Rui, and C. Z. Chen. 1985. Permian and Triassic sedimentary facies and paleogeography of South China. In K. Nakazawa and J. M. Dickins, eds., *The Tethys: Her Paleogeography and Paleobiogeography from Paleozoic to Mesozoic,* pp. 59–82. Tokyo: Tokai University Press.

Sherlock, R. L. 1948. *The Permo-Triassic Formations.* London: Hutchinson's Scientific & Techincal Publications.

Signor, P. W. 1990. The geologic history of diversity. *Annual Review of Ecology and Systematics* 21:509–539.

Signor, P. W., III, and J. H. Lipps. 1982. Sampling bias, gradual extinction patterns and catastrophes in the fossil record. In L. T. Silver and P. H. Schultz, eds., *Geological Implications of Impacts of Large Asteroids and Comets on the Earth,* pp. 291–296. Geological Society of America Special Paper No. 190. Boulder, Colo.: Geological Society of America.

Sigurdsson, H. 1990a. Assessment of the atmospheric impact of volcanic eruptions. In V.L. Sharpton and P. D. Ward, eds., *Global Catastrophes in Earth History,* pp. 99–110. Geological Society of America Special Paper No. 247. Boulder, Colo.: Geological Society of America.

Sigurdsson, H. 1990b. Evidence of volcanic loading of the atmosphere and climate response. *Palaeogeography, Palaeoclimatology, Palaeoecology (Global and Planetary Change Section)* 89:277–289.

Simberloff, D. S. 1974. Permo-Triassic extinctions: effects of area on biotic equilibrium. *Journal of Geology* 82:267–274.

Simpkin, T., and R. S. Fisk. 1983. *Krakatau 1883: The Volcanic Eruption and its Effects.* Washington, D.C.: Smithsonian Institution Press.

Skagami, S. 1981. Upper Permian Bryozoa from Guryul Ravine and the Spur three kilometers north of Barus. *Palaeontolgia Indica, NS* 46:45–64.

Sloan, C. L., J. C. G. Walker, T. C. Moore, Jr., D. K. Rea, and J. C. Zachos. 1992. Possible methane-induced polar warming in the early Eocene. *Nature* 357:320–322.

Sloan, R. 1985. Periodic extinctions and radiations of Permian terrestrial faunas and the rapid mamalization of therapsids. *Geological Society of America Abstracts with Program* 17(7):719.

Sloss, L. L. 1963. Sequences in the cratonic interior of North America. *Geological Society of American Bulletin* 74:93–113.

Sloss, L. L. 1992. Tectonic episodes of cratons: conflicting North American concepts. *Terra Nova* 4:320–328.

Smith, A. B. 1988. Late Palaeozoic biogeography of East Asia and palaeontological constraints on the plate tectonic reconstructions. *Philosophical Transactions of the Royal Society of London, Series A* 326:189–227.

Smith, A. B., and C. Patterson. 1988. The influence of taxonomic method on the perception of patterns of evolution. *Evolutionary Biology* 23:127–216.

Smith, A. G., A. M. Hurley, and J. C. Briden. 1981. *Phanerozoic Palaeocontiental World Maps.* Cambridge: Cambridge University Press.

313

REFERENCES

Smith, R. L. 1960. Ash-flow magmatism. In C. E. Chapin and W. E. Easton, eds., *Geological Society of America Special Paper* 180:5–27.

Sokratov, B. G. 1983. Oldest Triassic strata and the Permian-Triassic boundary in the Caucasus and Middle East. *International Geology Review* 25:483–496.

Southam, J. R., and W. W. Hay. 1981. Global sedimentary mass balance and sea level changes. In C. Emiliani, ed., *The Oceanic Lithosphere. The Sea*, vol. 7, pp. 1617–1684. New York: Wiley-Interscience.

Southam, J. R., W. H. Peterson, and G. W. Brass. 1982. Dynamics of anoxia. *Palaeogeography, Palaeoclimatology, Palaeoecology* 40:183–198.

Spitzy, A., and E. T. Degens. 1985. Modeling stable isotopic fluctuations through geologic time. *Mitteilungen Geologische-Paläontologich Institut Universitat Hamburg* 59:155–166.

Stafford, R. A. 1989. *Scientist of Empire*. Cambridge: Cambridge University Press.

Stanley, G. D., Jr. 1988. The history of early Mesozoic reef communites: a three-step process. *Palaios* 3:170–183.

Stanley, S. M. 1970. Relation of shell form to life habits in the Bivalvia. *Geological Society of America Memoir* 125:1–296.

Stanley, S. M. 1979. *Macroevolution*. San Francisco: W. H. Freeman.

Stanley, S. M. 1984. Marine mass extinctions: a dominant role for temperatures. In M. H. Nitecki, ed., *Extinctions*, pp. 69–117. Chicago: University of Chicago Press.

Stanley, S. M. 1988a. Climatic cooling and mass extinction of Paleozoic reef communities. *Palaios* 3:228–232.

Stanley, S. M. 1988b. Paleozoic mass extinctions: shared patterns suggest global cooling as a common cause. *American Journal of Science* 288:334–352.

Stanley, S. M. 1990. Delayed recovery and the spacing of major extinctions. *Paleobiology* 16:401–414.

Steiner, M., J. Ogg, Z. Zhang, and S. Sun. 1989. The Late Permian/Early Triassic magnetic polarity time scale and plate motions of South China. *Journal of Geophysical Research* 94(B6):7343–7363.

Stemmerik, L., and S. Piasecki. 1991. The Upper Permian of East Greenland— A review. *Zentralblatt für Geologie und Palaontologie* 1991(4):825–837.

Stepanov, D. L., F. Golshani, and J. Stocklin. 1969. Upper Permian and Permian-Triassic boundary in North Iran. *Geological Survey of Iran, Report* 12:72.

Stevens, C. H. 1977. Was development of brackish oceans a factor in Permian extinctions. *Geology* 88:133–138.

Stigler, S. M., and M. J. Wagner. 1987. Substantial bias in nonparametric tests for periodicity in geophysical data. *Science* 238:940–945.

Stille, H. 1924. *Grundfragen der verfleichenden Tektonik*. Berlin: Borntraeger.

Strathman, R. R. 1978. Progressive vacating of adaptive types during the Phanerozoic. *Evolution* 32:907–914.

REFERENCES

Strothers, R. B., and M. R. Rampino. 1990. Periodicity in flood basalts, mass extinctions and impacts; a statistical view and a model. In V. L. Sharpton and P. D. Ward, eds., *Global Catastrophes in Earth History*, pp. 9–17. Geological Society of America Special Paper 247. Boulder, Colo.: Geological Society of America.

Strothers, R. B., J. A. Wolf, S. Self, and M. R. Rampino. 1986. Basaltic fissure eruptions, plume heights and atmospheric aerosols. *Geophysical Research Letters* 13:725–728.

Stucky, R. K. 1987. Evolution of land mammal diversity in North America during the Cenozoic. In H. H. Genoways, ed., *Current Mammalogy*, vol. 2., pp. 375–432. New York: Plenum Press.

Suess, E. 1901. *Das Antlitz der Erde*. Prague: Tempsky and Freytag.

Sukhov, L. G., Y. A. Bespalaya, and D. A. Dolin. 1965. Biostratigraphy of volcanic formations in the western part of the Tunguska syneclise. *Doklady Akad. Nauk SSSR Geology Ser.* (Translated) 169:107–113.

Sukhov, L. G., and V. S. Golubkov. 1965. Principles of stratigraphic classification and correlation of ancient volcanic strata applied to the northwestern Siberian platform. *Doklady Akad. Nauk SSSR Geology Ser.* (Translated) 162:113–115.

Sun, Y. Y., Z. F. Chai, S. L. Ma, W. Y. Mao, D. Y. Xu, Q. W. Zhang, Z. Z. Yang, J. Z. Sheng, C. Z. Chen, L. Rui, X. L. Liang, and J. W. Hi. 1984. The discovery of iridium anomaly in the Permian-Triassic boundary clay in Changxing, Zhijing, China, and its significance. In G. Tu, ed., *Developments in Geosciences*, pp. 235–245. Beijing: Science Press.

Sweeney, J. F., J. R. Weber, and S. M. Blasco. 1982. Continental ridges in the Arctic Ocean: Lorex constraints. *Tectonophysics* 89:217–237.

Sweet, W. C. 1973. Late Permian and Early Triassic conodont faunas. In A. Logan and L. V. Hills, eds., *The Permian and Triassic Systems and Their Mutual Boundary*, Memoir 2, pp. 630–647. Calgary: Canadian Society of Petroleum Geologists.

Sweet, W. C. 1979. Graphic correlation of Permo-Triassic rocks in Kashmir, Pakistan and Iran. *Geologica et Palaeontologica* 13:239–248.

Sweet, W. C. 1992. A conodont-based high-resolution biostratigraphy for the Permo-Triassic boundary interval. In W. C. Sweet, Z. Y. Yang, J. M. Dickins, and H. F. Yin, eds., *Permo-Triassic Boundary Events in the Eastern Tethys*, pp. 120–133. Cambridge: Cambridge University Press.

Sweet, W. C., Z. Y. Yang, J. M. Dickins, and H. F. Yin. 1992. Permo-Triassic events in the eastern Tethus—an overview. In W. C. Sweet, Z. Y. Yang, J. M. Dickins, and H. F. Yin, eds., *Permo-Triassic Events in Eastern Tethys*, pp.1–8. Cambridge: Cambridge University Press.

Tappan, H. 1968. Primary production, isotopes, extinctions and the atmosphere. *Palaeogeography, Palaeoclimatology, Palaeoecology* 4:187–210.

Tappan, H. 1970. Phytoplankton abundance and late Paleozoic extinctions: a reply. *Palaeogeography, Palaeoclimatology, Palaeoecology* 8:56–66.

Tappan, H. 1982. Extinction or survival: selectivity and causes of Phanerozoic crises. In L. T. Silver and P. H. Schultz, eds., *Global Implications of Impacts of Large Asteroids and Comets on the Earth*, pp. 265–276. Geological Society of America Special Paper No. 247. Boulder, Colo.: Geological Society of America.

Tappan, H., and A. R. Loeblich. 1973. Smaller protistan evidence and explanation of the Permian-Triassic crisis. In A. Logan and L. V. Hills, eds., *The Permian and Triassic Systems and Their Mutual Boundary*, Memoir 2, pp. 465–480. Calgary: Canadian Society of Petroleum Geologists.

Tappan, H., and A. R. Loeblich. 1988. Foraminiferal evolution, diversification, and extinction. *Journal of Paleontology* 62:695–714.

Taraz, H. 1969. Permo-Triassic section in Central Iran. *American Association of Petrolem Geologists Bulletin* 53:688–693.

Taraz, H. 1971. Uppermost Permian and Permo-Triassic transition beds in Central Iran. *American Association of Petroleum Geologists* 55:1280–1294.

Taraz, H. 1973. Correlation of uppermost Permian in Iran, Central Asia and South China. *American Association of Petroleum Geologists Bulletin* 57:1117–1133.

Taylor, E. L., T. N. Taylor, and N. R. Cuneo. 1992. The present is not the key to the past: a polar forest from the Permian of Antarctica. *Science* 257:1675–1677.

Taylor, P. D., and G. P. Larwood. 1988. Mass extinctions and the pattern of bryozoan evolution. In G. P. Larwood, ed., *Extinction and the Fossil Record*, pp. 99–119. Oxford: Clarendon Press.

Teichert, C. 1990. The Permian-Triassic boundary revisited. In E. G. Kauffman and O. H. Walliser, eds., *Extinction Events in Earth History*, pp. 199–238. Berlin: Springer-Verlag.

Teichert, C., and B. Kummel. 1973. Permian-Triassic boundary in the Kap Stotch area, East Greenland. In A. Logan and L. V. Hills, eds., *The Permian and Triassic Systems and Their Mutual Boundary*, Memoir 2, pp. 269–285. Calgary: Canadian Society of Petroleum Geologists.

Teichert, C., and B. Kummel, B. 1976. Permian-Triassic boundary in the Kap Stosch area, East Greenland. *Meddelelser om Grønland* 597:1–54.

Teichert, C., B. Kummel, and H. M. Kapoor. 1970. Mixed Permian-Triassic fauna, Guryul Ravine, Kashmir. *Science* 167:174–175.

Teichert, C., B. Kummel, and W. Sweet. 1973. Permian-Triassic strata, Kuh-E-Ali Bashi, northwestern Iran. *Bulletin of the Museum of Comparative Zoology, Harvard* 145:359–472.

Thackeray, J. F., N. J. van der Merwe, J. A. Lee-Thorp, A. Sillen, J. L. Lanham, R. Smith, A. Keyser, and P. M. S. Monteiro. 1990. Changes in carbon isotope ratios in the late Permian recorded in therapsid tooth apatite. *Nature* 347:751–753.

Thierstein, H. R., and W. H. Berger. 1978. Injection events in ocean history. *Nature* 276:461–466.

Thompson, K. S. 1977. The pattern of diversification among fishes. In A. Hallam, ed., *Patterns of Evolution*, pp. 377–404. Amsterdam: Elsevier.

Tollmann, A., and E. Kristan-Tollmann. 1985. Paleogeography of the European Tethys from Paleozoic to Mesozoic and the Triassic relations of the eastern part of Tethys and Panthalassa. In K. Nakazawa and J. M. Dickins, eds., *The Tethys: Her Paleogeography and Paleobiogeography from Paleozoic to Mesozoic*, pp. 3–22. Toyko: Tokai University Press.

Tozer, E. T. 1979. The significance of the ammonoids *Paratirolites* and *Otoceras* in correlating the Permian-Triassic boundary beds of Iran and the People's Republic of China. *Canadian Journal of Earth Sciences* 16:1524–1532.

Tozer, E. T. 1980. Triassic Ammonoidea: classification, evolution and relationship with Permian and Jurassic forms. In M. R. House and J. R. Senior, eds., *The Ammonoidea*, Systematics Association special volume, pp. 66–100. London: Academic Press.

Tozer, E. T. 1984. The Trias and its ammonoids: evolution of a time scale. *Geological Society of Canada Miscellaneous Report* 35:171.

Tozer, E. T. 1988a. Definition of the Permian-Triassic (P-T) boundary: the question of the age of the *Otoceras* beds. *Memorie della Società Geologica Italiana* 34:291–301.

Tozer, E. T. 1988b. Towards a definition of the Permian-Triassic boundary. *Episodes* 11:251–255.

Traverse, A. 1988. Plant evolution dances to a different beat. Plant and animal evolutionary mechanisms compared. *Historical Biology* 1:277–302.

Turcotte, D. L., and K. Burke. 1978. Global sea-level changes and the thermal structure of the earth. *Earth and Planetary Science Letters* 41:341–346.

Umbgrove, J. H. F. 1947. *The Pulse of The Earth*. The Hague: Martinus Nijhoff.

Vail, P. R., and R. M. Mitchum. 1979. Global cycles and sea-level change and their role in exploration. *Proceedings of the 10th World Petroleum Congress, vol. 2, Exploration Demand and Supply*, pp. 95–104. Bucharest, Romania.

Vail, P. R., R. M. Mitchum, and S. Thompson III. 1977. Seismic stratigraphy and global changes of sea level, part 4: global cycles of relative changes of sea level. *American Association of Petroleum Geologists Memoir* 26:83–97.

Valentine, J. W. 1969. Patterns of taxonomic and ecological structure of the shelf benthos during Phanerozoic time. *Palaeontology* 12:684–709.

Valentine, J. W. 1971. Resource supply and species diversity patterns. *Lethaia* 4:51–61.

Valentine, J. W. 1973a. *Evolutionary Paleoecology of the Marine Biosphere*. Englewood Cliffs, N.J.: Prentice-Hall.

Valentine, J. W. 1973b. Plates and provinciality, a theoretical history of environmental discontinuities. In N. F. Hughes, ed., *Organisms and Continents Through Time*. Special Papers in Palaeontology, no. 12, pp. 79–92.

REFERENCES

Valentine, J. W. 1980. Determinants of diversity in higher taxonomic categories. *Paleobiology* 6:444–450.

Valentine, J. W. 1986. The Permian-Triassic extinction event and invertebrate developmental models. *Bulletin of Marine Science* 39:607–615.

Valentine, J. W. 1989. How good was the fossil record? Clues from the Californian Pleistocene. *Paleobiology* 15:83–94.

Valentine, J. W., and D. Jablonski. 1983. Larval adaptations and patterns of brachiopod diversity in space and time. *Evolution* 37:1052–1061.

Valentine, J. W., and D. Jablonski. 1986. Mass extinctions: sensitivity of marine larval types. *Proceedings of the National Academy of Sciences USA* 83:6912–6914.

Valentine, J. W., and D. Jablonski. 1991. Biotic effects of sea level change: the Pleistocene test. *Journal of Geophysical Research* 96:6873–6878.

Valentine, J. W., and E. M. Moores. 1970. Plate-Tectonic regulation of faunal diversity and sea level: a model. *Nature* 228:657–659.

Valentine, J. W., and E. M. Moores. 1972. Global tectonics and the fossil record. *Journal of Geology* 80:167–184.

Valentine, J. W., and E. M. Moores. 1973. Provinciality and diversity across the Permian-Triassic boundary. In A. Logan and L. V. Hills, eds., *The Permian and Triassic Systems and Their Mutual Boundary*, Memoir 2, pp. 759–766. Calgary: Canadian Society of Petroleum Geologists.

Van Valen, L. M. 1984. A resetting of Phanerozoic community evolution. *Nature* 307:50–52.

Van der Voo, R., and R. B. French. 1974. Apparent polar wandering for the Atlantic-bordering continents: Late Carboniferous to Eocene. *Earth-Science Reviews* 10:99–119.

Van der Voo, R., J. Peinado, and C. R. Scotese. 1984. A paleomagnetic reevaluation of Pangea reconstructions. *American Geophysical Union Geodynamics Series* 12:11–26.

Veevers, J. J. 1989. Middle/Late Triassic (230 ± 5 Ma) singularity in the stratigraphic and magmatic history of the Pangean heat anomaly. *Geology* 17:784–787.

Veevers, J. J., and C. M. Powell. 1987. Late Paleozoic glacial episodes in Gondwanaland reflected in transgressive-regressive depositional sequences in Euramerica. *Geological Society of America Bulletin* 98:475–487.

Veizer, J. 1989. Strontium isotopes in seawater through time. *Annual Review of Earth and Planetary Sciences* 17:141–167.

Vermeij, G. J. 1977. The Mesozoic marine revolution: evidence from snails, predators and grazers. *Paleobiology* 3:245–258.

Vermeij, G. J. 1987. *Evolution and Escalation*. Princeton: Princeton University Press.

Visscher, H., and W. A. Brugman. 1988. The Permian-Triassic boundary in the Southern Alps: a palynological approach. *Memorie della Società Geologica Italiana* 34:121–128.

Vogt, P. R. 1972. Evidence for global synchronism in mantle plume convection, and possible significance for geology. *Nature* 240:338–342.

Vogt, P. R. 1975. Changes in geomagnetic reversals frequency at times of tectonic change: evidence for coupling between core and upper mantle processes. *Earth and Planetary Science Letters* 25:313–321.

Vogt, P. R. 1979. Global magmatic episodes: new evidence and implications for the steady-state mid-ocean ridge. *Geology* 7:93–98.

Vogt, P. R. 1989. Volcanogenic upwelling of anoxic, nutrient-rich water: a possible factor in carbonate-bank/reef demise and benthic faunal extinctions? *Geological Society of America Bulletin* 101:1225–1245.

Waagen, W. 1891. Salt Range fossils. IV. Geological results. *Palaeontological Indica Ser. 13* 4:89–242.

Wadia, D. N. 1957. *Geology of India.* 3d ed. London: Macmillan.

Walker, T. D. 1985. Diversification functions and the rate of taxonomic evolution. In J. W. Valentine, ed., *Phanerozoic Diversity Patterns*, pp. 311–334. Princeton: Princeton University Press.

Walsh, J. J. 1991. Importance of continental margins in the marine biogeochemical cycling of carbon and nitrogen. *Nature* 350:53–55.

Wang, K. 1992. Glassy microspherules (microtektites) from an Upper Devonian limestone. *Science* 256:1547–1550.

Wang, K., B. D. E. Chatterton, M. Attrep, Jr., and C. J. Orth. 1992. Iridium abundance maxima at the latest Ordovician mass extinction horizon Yangtze Basin, China: terrestrial or extraterrestrial? *Geology* 20:39–42.

Wardlaw, B., and R. E. Grant. 1992. Personal communication.

Waterhouse, J. B. 1973. The Permian-Triassic boundary in New Zealand and New Caledonia and its relationship to world climatic changes and extinction of Permian life. In A. Logan and L. V. Hills, eds., *The Permian and Triassic Systems and Their Mutual Boundary*, Memoir 2, pp. 445–464. Calgary: Canadian Society of Petroleum Geologists.

Waterhouse, J. B. 1978. Chronostratigraphy for the world Permian. *AAPG Studies in Geology* 6:299–322.

Waterhouse, J. B., and G. R. Shi. 1990. Late Permian brachiopod faunules from the Marsyangdi Formation, north central Nepal, and the implications for the Permian-Triassic boundary. In D. L. MacKinnon, D. E. Lee, and J. D. Campbell, eds., *Brachiopods Through Time*, pp. 381–387. Rotterdam: Balkema.

Watson, M. P., A. B. Hayward, D. N. Parkinson, and Z. M. Zhang. 1987. Plate tectonic history, basin development and petroleum source rock deposition onshore China. *Marine and Petroleum Geology* 4:205–225.

Waugh, B. 1973. The distribution and formation of Permian-Triassic red beds. In A. Logan and L. V. Hills, eds., *The Permian and Triassic Systems and Their Mutual Boundary*, Memoir 2, pp. 678–693. Calgary: Canadian Society of Petroleum Geologists.

REFERENCES

Wiedmann, J. 1973. Evolution or revolution of ammonoids at Mesozoic system boundaries. *Biological Reviews* 48:159–194.

Wignall, P. B., and A. Hallam. 1992. Anoxia as a cause of the Permian/Triassic extinction: facies evidence from northern Italy and the western United States. *Palaeogeography, Palaeoclimatology, Palaeoecology.* 93:21–46.

Wilde, P., and W. B. N. Berry. 1984. Destabilization of the oceanic density structure and its significance to marine "extinction" events. *Palaeogeography, Palaeoclimatology, Palaeoecology* 48:143–162.

Wilde, P., and W. B. N. Berry. 1986. The role of oceanographic factors in the generation of global bio-events. In O. Walliser, ed., *Global Bio-Events*, pp. 75–91. Berlin: Springer-Verlag.

Wilgus, C. K., B. S. Hastings, C. G. St. C. Kendall, H. W. Posamentier, C. A. Ross, and J. C. von Wagoner. 1988. *Sea-Level Changes: An Integrated Approach.* Tulsa: Society of Economic Paleontologists and Mineralogists Special Publication no. 42.

Williamson, M. 1988. Relationship of species number to area, distance and other variables. In A. A. Myers and P. S. Giller, eds., *Analytical Biogeography*, pp. 91–115. London: Chapman and Hall.

Wilson, J. L. 1975. *Carbonate Facies in Geologic History.* New York: Springer-Verlag.

Wilson, J. T. 1966. Did the Atlantic close and then re-open? *Nature* 211:676–681.

Wise, K. P., and T. J. M. Schopf. 1981. Was marine faunal diversity in the Pleistocene affected by changes in sea level? *Paleobiology* 7:394–399.

Wootton, R. J. 1990. Major insect radiations. In P. D. Taylor and G. P. Larwood, eds. *Major Evolutionary Radiations*, pp. 187–208. Oxford: Clarendon Press.

Worsley, T. R., and R. D. Nance. 1989. Carbon redox and climate control through earth history: a speculative reconstruction. *Palaeogeography, Palaeoclimatology, Palaeoecology (Global and Planetary Change Section)* 75:259–282.

Worsley, T. R., and D. L. Kidder. 1991. First-order coupling of paleogeography and CO_2 with global surface temperature and its latitudinal contrast. *Geology* 19:1161–1164.

Worsley, T. R., J. B. Moody, and R. D. Nance. 1985. Proterozoic to Recent tectonic tuning of biogeochemical cycles. In E. T. Sundquist and W. S. Broecker, eds., *The Carbon Cycle and Atmospheric CO_2: Natural Variations Archean to Present*, pp. 561–572. Washington, D.C.: American Geophysical Union.

Worsley, T. R., R. D. Nance, and J. B. Moody. 1984. Global eustacy for the past 2 billion years. *Marine Geology* 58:373–400.

Worsley, T. R., R. D. Nance, and J. B. Moody. 1986. Tectonic cycles and the history of the earth's biogeochemical and paleoceanographic record. *Paleoceanography* 1:233–263.

320

Wright, J. 1989. REE in fossil apatite record Phanerozoic OAE. *28th International Geologic Congress Abstracts* 3:383–384.

Wright, J., H. Schrader, and W. T. Holser. 1987. Paleoredox variations in ancient oceans recorded by rare earth elements in fossil apatite. *Geochimica et Cosmochimica Acta* 5:632–644.

Wyatt, A. R. 1984. Relationship between continental area and elevation. *Nature* 311:370–372.

Wyatt, A. R. 1987. Shallow water areas in space and time. *Journal of Geological Society of London* 144:115–120.

Wynne, A. B. 1878. On the geology of the Salt Range in the Punjab. *Geological Survey of India, Memoir* 14:1–313.

Xu, D. Y., S. L. Ma, Z. F. Chai, X. Y. Mao, Y. Y. Sun, Q. W. Zhang, and Z. Z. Yang. 1985. Abundance variation of iridium and trace elements at the Permian-Triassic boundary at Shangsi in China. *Nature* 314:154–156.

Xu, D. Y., Q. W. Zhang, and Y. Y. Sun. 1988. Mass extinction—a fundamental indicator for major natural divisions of geological history. *Acta Geologica Sinica* 1:1–15.

Xu, D. Y., Q. W. Zhang, Y. Y. Sun, Z. Yan, Z. F. Chai, and J. W. He. 1989. *Astrogeological Events in China*. New York: Van Nostrand Reinhold.

Xu, G. 1991. Stratigraphical time-correlation and mass extinction event near Permian-Triassic boundary in South China. *Journal of China University of Geosciences* 2:36–46.

Yang, W. R., and N. Y. Jiang. 1981. On the depositional characters and microfacies of the Changhsing Formation and the Permo-Triassic boundary in Changxing, Zhejiang. *Bulletin of the Nanjing Institute of Geology & Palaeontology, Academica Sinica* 2:129–131.

Yang, Z. Y., and Z. S. Li. 1992. Permo-Triassic boundary relations in south China. In W. C. Sweet, Z. Y. Yang, J. M. Dickins, and H. F. Yin, eds., *Permo-Triassic Events in the Eastern Tethys*, pp. 9–20. Cambridge: Cambridge University Press.

Yang, Z. Y. and H. F. Yin. 1987. Achievements in the study of Permo-Triassic events in South China. *Advances in Science of China, Earth Sciences* 2:23–43.

Yang, Z. Y., H. F. Yin, S. B. Wu, F. Q. Yang, M. H. Ding, and G. R. Xu. 1987. Permian-Triassic boundary. Stratigraphy and faunas of South China. *Ministry of Geology and Mineral Resources, Geological Memoir, Series 2* 6:1–379.

Yao, Z. Q. and O. Y. Shu. 1980. On the Paleophyte-Mesophyte boundary. *Papers for the 5th International Palynological Conference. Nanjing Institute of Geology and Palaeontology.*

Yin, H. F. 1985a. Bivalves near the Permian-Triassic boundary in South China. *Journal of Paleontology* 59:572–600.

Yin, H. F. 1985b. On the transitional bed and the Permian-Triassic boundary in South China. *Newsletter on Stratigraphy* 15(1):13–27.

REFERENCES

Yin, H. F., S. Huang, K. X. Zhang, H. J. Hansen, F. Q. Yang, M. H. Ding, and X. M. Die. 1992. The effects of volcanism on the Permo-Triassic mass extinction in South China. In W. C. Sweet, Z. Y. Yang, J. M. Dickins, and H. F. Yin, eds., *Permo-Triassic Events in the Eastern Tethys*, pp. 146–157. Cambridge: Cambridge University Press.

Yin, H. F., S. Huang, K. X. Zhang, F. Q. Yang, M. H. Ding, B. I. Zianmei, and Z. Suxian. 1989. Volcanism at the Permian-Triassic boundary in South China and its effects on mass extinction. *Acta Geologica Sinica* 2:417–431.

Yin, H. F., G. R. Xu, and M. H. Ding. 1984. Palaeozoic-Mesozoic alternation of marine biota in South China. *Scientific Papers on Geology for International Exchange—Prepared for the 27th International Geological Congress*, 195–204.

Yin, H. F., G. Q. Yang, K. X. Zhang, and W. P. Yang. 1988. A proposal to the biostratigraphic criteron of Permian/Triassic boundary. *Memorie della Società Geologica Italiana* 34:329–344.

Zakharov, Y. D. 1992. The Permo-Triassic boundary in the southern and eastern USSR and its international correlation. In W. C. Sweet, Z. Y. Yang, J. M. Dickins, and H. F. Yin, eds., *Permo-Triassic Events in the Eastern Tethys*, pp. 46–55. Cambridge: Cambridge University Press.

Zakharov, Y. D., and A. N. Sokarev. 1991. Permian-Triassic paleomagnetism of Eurasia. *Proceedings of Shallow Tethys* 3:313–323.

Zhao, J. K., J. Z. Sheng, Z. Q. Yao, X. L. Liang, C. Z. Chen, L. Rui, and A. T. Liao. 1981. The Changhsingian and Permian-Triassic boundary of South China. *Bulletin of the Nanjing Institute of Geology and Palaeontology, Academica Sinica* 2:1–112.

Zharkov, M. A. 1981. *History of Paleozoic Salt Accumulation*. Berlin: Springer Verlag.

Zhou, L., and F. T. Kyte. 1988. The Permian-Triassic boundary event: a geochemical study of three Chinese sections. *Earth and Planetary Science Letters* 90:411–421.

Ziegler, A. M. 1990. Phytogeographic patterns and continental configurations during the Permian Period. In W. S. McKerrow and C. R. Scotese, eds., *Palaeozoic Palaeogeography and Biogeography*, pp. 363–379. London: Geologica Society Memoir.

Zolotukhin, V. V., and A. I. Al'Mukhamedov. 1988. Traps of the Siberian platform. In J. D. Macdougall, ed., *Continental Flood Basalts*, pp. 273–310. Dordrecht, Holland: Kluwer.

INDEX

Text: 9.5/12 Trump Mediaeval
Compositor: Impressions, a division of Edwards Brothers
Printer: Edwards Brothers
Binder: Edwards Brothers